HUMAN LONGEVITY

Omega-3 Fatty Acids, Bioenergetics, Molecular Biology, and Evolution

HUMAN LONGEVITY

Omega-3 Fatty Acids, Bioenergetics,
Molecular Biology, and Evolution

RAYMOND C. VALENTINE
DAVID L. VALENTINE

CRC Press
Taylor & Francis Group
Boca Raton London New York

CRC Press is an imprint of the
Taylor & Francis Group, an **informa** business

First published in paperback 2024

First published 2015
by CRC Press
2385 NW Executive Center Drive, Suite 320, Boca Raton FL 33431

and by CRC Press
4 Park Square, Milton Park, Abingdon, Oxon, OX14 4RN

First issued in hardback 2019

CRC Press is an imprint of Taylor & Francis Group, LLC

© 2015, 2019, 2024 Taylor & Francis Group, LLC

Library of Congress Cataloging-in-Publication Data

Valentine, R. C. (Raymond Carlyle), 1936-
    Human longevity : omega-3 fatty acids, bioenergetics, molecular biology, and evolution / Raymond C. Valentine and David L. Valentine.
        pages cm
    Includes bibliographical references and index.
    ISBN 978-1-4665-9486-9 (hardcover)
    1. Aging--Genetic aspects. 2. Mitochondria. 3. Oxidative stress. 4. Omega-3 fatty acids--Health aspects. I. Valentine, David L. II. Title.

QP86.V35 2014
612.6'7--dc23   2014010535

ISBN: 978-1-4665-9486-9 (hbk)
ISBN: 978-1-03-292478-6 (pbk)
ISBN: 978-0-429-16877-2 (ebk)

DOI: 10.1201/b17458

Visit the Taylor & Francis Web site at
http://www.taylorandfrancis.com

and the CRC Press Web site at
http://www.crcpress.com

Contents

Preface...xiii
Acknowledgments..xv
About the Authors..xvii

SECTION I   Introduction to the Science of Human Aging

Chapter 1   Mitochondrial Hypothesis of Aging Is Undergoing Revision..............3

1.1   Historical Perspective...3
1.2   Conventional Mitochondrial Theory of Aging Is
         Undergoing Revision 4
1.3   Changes Have Come to the Famous Oxidative Stress or
         Free Radical Theory of Aging..5
1.4   Membrane Polyunsaturation Theory Seems to Link
         Mitochondrial-Oxidative Theories of Aging 6
1.5   Summary 8
References 9

Chapter 2   Oxidative Stress Defined as a Deadly Free Radical-Mediated
Chain Reaction: Case History of Paraquat.. 11

2.1   Paraquat Generates Free Radicals That Target and
         Destroy Chloroplasts along with the Plant's Basic Energy
         Supply 12
2.2   Nonlethal Doses of Paraquat and Related Chemicals Act
         as Neurotoxins and Selectively Kill Parkinsonian Neurons ... 15
2.3   Death and Murder by Paraquat... 16
2.4   Summary 17
References 17

Chapter 3   Membranes of Deep-Sea Bacteria as Surrogates for
Mitochondrial Membranes of Humans 19

3.1   Oxygen-Damaged Membrane Fatty Acids Score Positive
         as Mutagens in the Ames Carcinogen Test 20
3.2   Lessons from DHA/EPA-Producing Bacteria..........................21

3.3   Tripartite Membrane Fatty Acid Blending Code Can
       Explain Bizarre Properties of an EPA Recombinant of
       E. coli..24
3.4   The DHA Principle Has a Wide Range of Applications.........26
3.5   Tripartite Membrane Fatty Acid Blending Code Is Also
       Likely Universal 27
3.6   Applications to Mitochondria...29
3.7   Bacterial Recombinants and Mutants Deserve More
       Attention as Models for Mitochondrial Membranes 31
3.8   Summary 32
References 32

SECTION II   Darwinian Selection of Membranes Enabling Longevity

Chapter 4   Protective Mechanisms for EPA Membranes in *C. elegans* and
               Their Relationship to Life Span 37

4.1   Feeding an Excess of EPA Decreases Life Span of
       C. elegans 37
4.2   *C. elegans* Synthesizes One of the Most Highly
       Unsaturated Membranes in Nature..39
4.3   EPA Benefits *C. elegans* by Enabling Extreme
       Membrane Motion Important for Rapid Cycling of the
       Synaptic Vesicles Essential for Fast Firing of Bending
       Muscles 40
4.4   Sensory Perception in *C. elegans* Likely Depends
       on EPA Enabling Extraordinary Membrane Motion...............42
4.5   Sensory Systems That Protect against Membrane
       Oxidation and Oxidative Stress..43
4.6   An Extremely Long-Lived Mutant Decreases EPA While
       Increasing Levels of Monounsaturated Chains 45
4.7   *C. elegans* Targets EPA Away from Mitochondrial
       Membranes Likely as a Protective Mechanism against
       Oxidative Damage 47
4.8   Growth of *C. elegans* Requires Methyl-Branched Fatty
       Acids, Which Are Peroxidation Resistant 48
4.9   Mutational Analysis of Nuclear Hormone Receptors
       Provides an Alternative View of the Roles of Lipids in
       Regulating Life Span...48
4.10   Summary 50
References 51

Chapter 5 Remarkable Longevity of Queens of Social Insects Likely
Involves Dietary Manipulation to Minimize Levels of
Polyunsaturates and Decrease Membrane Peroxidation 53

    5.1   Royalactin 54
    5.2   Royal Jelly 54
    5.3   Hydroxy Fatty Acids Acting as Signaling Molecules
          Might Be Linked to Longevity of Queens 56
    5.4   Polyunsaturated Fatty Acids Missing from Royal Jelly
          May Be a Secret to Longevity of Queens 56
    5.5   Dietary Manipulation of Protein Levels Also Has
          Dramatic Effects on Longevity of Ants 57
    5.6   Lessons from *Drosophila* 59
    5.7   dFOXO in *Drosophila* 60
    5.8   Summary 60
    References 61

Chapter 6 Membrane Peroxidation Hypothesis Helps Explain Longevity
in Birds, Rodents, and Whales 63

    6.1   Extreme Flight of Hummingbirds Is Dependent on
          Highly Unsaturated Mitochondrial Membranes That
          Might Dictate Their Short Life Span 63
    6.2   Similar Brain DHA Levels in Long-Lived Pigeons versus
          Short-Lived Rats Seem to Defy Polyunsaturation Theory
          of Aging 66
    6.3   Naked Mole Rats Have One-Ninth the Level of DHA as
          a Mouse and Live Nine Times Longer 67
    6.4   Membrane Polyunsaturation Model Predicts Whales as
          the Longest-Lived Mammals 70
    6.5   Summary 73
    References 73

Chapter 7 Did Longevity Help Humans Become Super Humans? 75

    7.1   Caspari's Hypothesis of the Rise of Grandparents 75
    7.2   Fossil Teeth Show Longevity Occurred Late in Human
          Evolution 76
    7.3   Ancient Artifacts Show Cultural Revolution and
          Population Growth Coinciding with Longevity 77
    7.4   Does Bioenergetics Help Explain the Evolution of
          Grandparents? 77
    7.5   Evolution of Human Longevity as a Genetic Event 78
    7.6   Convergent Evolution of Longevity 79
    7.7   Present-Day Risks of Human Longevity 80

7.8   Revised Holy Grail of Aging..81
7.9   Summary 82
References 82

SECTION III   Revised Mitochondrial Membrane Hypothesis of Aging

Chapter 8   Mitochondrial Diseases and Aging Have Much in Common 85

8.1   Leber's Hereditary Optic Neuropathy (LHON) 85
8.2   Barth's Syndrome 88
8.3   Latent Mitochondrial Diseases Caused by Mutations
        in the mtDNA-Replicating Machine (POLG) May Not
        Display Symptoms for Up to Sixty Years...............................89
8.4   Friedreich's Ataxia 91
8.5   Summary 93
References 93

Chapter 9   Revised Mitochondrial Hypothesis of Aging Highlights Energy
                Deficiency Caused by Errors of Replication (Mutations) of mtDNA...95

9.1   MtDNA Encodes Seven Subunits of Complex 1 95
9.2   Mutated POLG in Mice Accelerates Aging 97
9.3   Mutator Genes Accelerate Neurodegeneration
        in Humans, Suggesting That the Brain Can Set the Pace
        of Aging...98
9.4   Recent Data Confirm the Mutator Concept and Help
        Explain How Mitochondrial Fusion Can Push Back
        against Aging...99
9.5   Mixing of Components during Mitochondrial
        Fusion Might Protect Membranes against Age-
        Dependent Damage 101
9.6   Summary 103
References 103

Chapter 10 Benefits of Polyunsaturated Mitochondrial Membranes................. 107

10.1   Conformational Dynamics Explain How Polyunsaturated
        Chains Are Harnessed for Bioenergetic Gain 107
10.2   Overexpressing DHA/EPA in Transgenic Mice Increases
        Rates of Electron Transport... 109
10.3   Mitochondrial Membrane Composition Reflects the Need
        to Balance Energy Production and Energy Conservation 109

10.4   Summary 113
References 113

Chapter 11 Mitochondrial Membranes as a Source of Reactive Oxygen
Species (ROS) 115

11.1   The Beneficial Role of ROS Production by Phagocytic
Cells Is to Kill Pathogens by Inflicting Catastrophic
Oxidative Damage, But There Are Side Effects 115
11.2   Mitochondria as a Major Source of ROS 118
11.3   Mitochondrial Polyunsaturated Membranes as a Major
Source of ROS 121
11.4   Summary 122
References 123

Chapter 12 Mitochondrial Membranes as Major Targets of Oxidation 125

12.1   Rhodopsin Membrane Disks Are Highly Enriched with
DHA and Age Rapidly 125
12.2   Lessons from DHA-Enriched Tails of Sperm 129
12.3   Mitochondrial Membranes and Their Proteins as Targets
of Oxidative Damage 131
12.4   Summary 134
References 134

SECTION IV   Many Mechanisms Have Evolved to Protect Human Mitochondrial Membranes, Enabling Longevity

Chapter 13 Apoptosis Caused by Oxidatively Truncated Phospholipids Can
Be Reversed by Several Mechanisms, Especially Enzymatic
Detoxification 139

13.1   Lipid Whiskers and Their General Properties 139
13.2   Lipid Whiskers Signal Phagocytic Cells to Converge on a
Damaged Membrane Site 141
13.3   Phagocytic Cells Hyperactivated by Lipid
Whiskers Can Increase Oxidative Damage Fostering
Inflammation 142
13.4   Oxidatively Truncated Phospholipids as Triggers of
Apoptosis Mediated by Mitochondria 143
13.5   Summary 144
References 145

Chapter 14 Selective Targeting of HUFAs Away from Cardiolipin and
Beta-Oxidation Combine to Protect Mitochondrial Membranes
against Oxidative Damage 147

14.1 Selective Fatty Acid Targeting Has Been Demonstrated
for DHA in Sperm and Other Cells...................................... 148
14.2 Selective Targeting and Avoidance of Polyunsaturated
Chains in Cardiolipin as a Pro-Longevity Mechanism......... 150
14.3 Beta-Oxidation Is Responsible for Degrading a Vast
Majority of DHA in the Body, Thus Minimizing DHA
Incorporation into Most Cellular Membranes....................... 151
14.4 Comparative Biochemistry of DHA Detoxification.............. 153
14.5 Summary 154
References 155

Chapter 15 Oxygen Limitation Protects Mitochondrial Phospholipids,
Especially Cardiolipin.. 157

15.1 DHA Turnover in the Brain Is Surprisingly Slow,
Suggesting the Operation of Novel Protective Mechanisms ... 157
15.2 O_2 Avoidance by Mitochondria 159
15.3 During O_2-Limited Conditions Cells Have the Option of
Rebalancing the Ratio of Respiration to Glycolysis: Case
History of Sperm 161
15.4 Model for O_2 Protection by Myelin Found in Root
Nodule Bacteria 163
15.5 Summary 164
References 165

Chapter 16 Uncoupling Proteins (UCPs) of Mitochondria Purposely Waste
Energy to Prevent Membrane Damage 167

16.1 Nature of Mitochondrial Energy UCPs and Their
Activation by PUFAs, HUFAs, and Fatty Acid
Peroxidation Products.. 167
16.2 A Molecular Model Linking Membrane Unsaturation
with Longevity... 171
16.3 Summary 172
References 173

Chapter 17 Mitochondrial Fission Protects against Oxidative Stress by
Minting a Continuous Supply of Cardiolipin and Other
Polyunsaturated Phospholipids 175

17.1 Discovery of MAMs... 175
17.2 Major Molecular Species of Mitochondrial Phospholipids
and Their Biosynthesis 177

17.3   Excessive Mitochondrial Fission May Generate Toxic
Mitochondria 179
17.4   Summary 181
References 182

Chapter 18   Mitophagy Eliminates Toxic Mitochondria 183

18.1   Brief Description of Mitophagy 183
18.2   Mitophagy in Yeast: A Mechanism to Decrease
Oxidative Stress.. 183
18.3   Roles of Mitophagy in Mice... 185
18.4   Summary 187
References 187

Chapter 19   Longevity Genes Likely Protect Membranes................................... 189

19.1   Ashkenazi Jewish Centenarians 189
19.2   FOXO Transcription Factors Govern Longevity in Model
Animals and Perhaps Humans 190
19.3   Hypothesis That Neurons Act as Master Regulators
of Aging and as a Source of Longevity Genes, Including
Apolipoprotein D (ApoD).. 194
19.4   Genetic Variants of Uncoupling Proteins (UCPs) as
Longevity Genes.. 197
19.5   Summary 197
References 198

Chapter 20   Aging as a Cardiolipin Disease That Can Be Treated 201

20.1   Working Definition of Cardiolipin Diseases 201
20.2   Definitive Proof That Double Bonds of Polyunsaturated
Membrane Phospholipids Can Cause Oxidative Death of
Cells.. 202
20.3   Deuterated 18:2 (D-18:2) Protects Neurons against
Oxidative Death in a Mouse Model of Parkinson's Disease ... 203
20.4   New Generation of Antioxidants Targeting
Mitochondrial Membranes Suppress Oxidative Damage
and Improve Mitochondrial Function in a Mouse Model
of Huntington's Disease (HD) 205
20.5   Cardiolipin Oxidation Mediates Neuron Death during
Traumatic Brain Injury in Rats, and Mitochondria-
Targeted Antioxidant XJB-5-131 Protects Cardiolipin
and Prevents Apoptosis 207
20.6   Summary 210
References 212

Index... 215

Preface

This book is built on the proposition that we age as our mitochondria age. We suggest a revised version of Harman's famous hypothesis featuring mitochondrial oxidative and energy stresses as the root causes of aging. It is well known that cellular death or apoptosis is triggered by energy stress or oxidative stress. There are convincing data showing that as mitochondria age, mitochondrial DNA (mtDNA) accumulates mutations, decreasing energy output. Protecting highly unsaturated mitochondrial membranes against oxidative stress also consumes an extraordinary amount of energy. Thus, an aging cell faces a double whammy of energy stress: decreasing energy production and high energy demand to maintain mitochondrial membrane integrity. The net effect is that as cells age, an increasing percentage of energy output from respiration is used for membrane maintenance. The challenge of this book is to develop a unified concept of aging that explains how mitochondrial membranes act as a tipping point, causing energy stress and eventually massive cellular death.

Acknowledgments

Once again, we thank our sister publisher, Garland Press, and the editors of *Molecular Biology of the Cell* for permission to use several figures and illustrations. Thanks to Hilary Rowe, our editor at Taylor & Francis/CRC Press, for advice, help, and encouragement. Thanks to Josie Banks-Kyle of CRC Press for help with permissions, illustrations, and coordinating the production of this book.

Special thanks to Cindy Anders for translating RCV's "hen tracks" scribbled in longhand on numerous yellow legal pads to the computer, editing the English, checking references, and obtaining permissions for illustrations, to name a few of her indispensable contributions. Thanks to Dr. Nicki Neff for helpful suggestions that improved the manuscript.

Raymond C. Valentine
Professor Emeritus, University of California at Davis

David L. Valentine
Professor, Department of Earth Science and Marine Science Institute
University of California at Santa Barbara

About the Authors

Raymond C. Valentine is currently professor emeritus at the University of California, Davis and visiting scholar in the Marine Science Institute at the University of California, Santa Barbara. He was also the scientific founder of Calgene, Inc. (Davis, California), now a campus of Monsanto, Inc. The author's scientific interests involve the use of reductionism to address problems of fundamental scientific and societal importance, such as agricultural productivity and aging. Some of his scientific accomplishments include the discovery of ferredoxin, the identification and naming of the nitrogen fixation (*nif*) genes, and the development of Roundup® resistance in crops. He holds BS and PhD degrees from the University of Illinois at Urbana-Champaign.

David L. Valentine is currently professor of earth science with affiliations in ecology, evolution, and marine biology, as well as the Marine Science Institute, at the University of California, Santa Barbara. The author's scientific interests involve the use of a systems-based approach to investigate the interaction between microbes and the earth, particularly in the subsurface and oceanic realms. He is best known for his research on the biogeochemistry of methane and other hydrocarbons, his works on archaeal metabolism and ecology, and his scientific work on the *Deepwater Horizon* oil spill. DLV holds BS and MS degrees from the University of California at San Diego and MS and PhD degrees from the University of California at Irvine.

Section I

Introduction to the Science of Human Aging

This book is dedicated to 486 researchers at 302 institutions in 50 countries who spent five years developing the recently published Global Burden of Disease Study. This study tracked trends in world health and seems to identify a critical tipping point occurring between 1990 and 2010. The research is detailed in seven papers published in the medical journal *Lancet*, which for the first time devoted an entire issue to one research study. Christopher Murray, who coordinated this research, summarizes the data as follows: "We are in a transition to a world where disability is the dominant concern as opposed to premature death. The pace of change is such that we are ill-prepared to deal with what the burden of disease is now in most places."

The Global Burden of Disease Study stands out as being the most detailed analysis of health of world populations ever attempted. Causes of death were charted for 235 diseases or conditions. Clearly some major changes have occurred over the twenty-year time frame of this study, and the data predict a trajectory for future world health. The major trend is the rise in age-dependent diseases. In essence, more people are surviving to die of diseases that occur primarily in the elderly. These include Alzheimer's disease, deaths from which tripled from 1990 to 2010, and Parkinson's disease, which doubled during this period.

Aging results in many nonlethal conditions that together place a great deal of pressure on the healthcare systems. These ailments include vision loss, hearing loss, and musculoskeletal pain. As the global population ages, the time of bad health tends to increase and is becoming perhaps the major challenge in world healthcare. Mental illness (including addiction) is now responsible for 23 percent of "years lived with disability," with low-back pain responsible for 11 percent. Overall, these data highlight future trends in world health driven by aging, and their purpose is to give governments, international agencies, donors, and researchers a new perspective from which to plan. The Bill and Melinda Gates Foundation provided $12 million for this $20 million project.

Numerous hypotheses attempting to unify mechanisms of aging have been proposed. One called the mitochondrial hypothesis of aging is widely quoted in this field. Mitochondria are powerhouses of our cells, and the mitochondrial theory of aging states that as energy production by mitochondria wanes with age, we age. Toxic by-products of mitochondrial respiration called free radicals or reactive oxygen species (ROS) are proposed to continuously damage mitochondria and cause aging. It is clear that human cells are genetically programmed to commit suicide when energy output by mitochondria is overwhelmed by energy demand.

We suggest that two major energy-related events occur in mitochondria as they age, eventually resulting in massive cellular death. The first is declining energy output by mitochondria, bringing cells closer and closer to a critical threshold of energy supply and triggering a program of cellular death. The second mechanism, and the main subject of this book, is that the energy cost to protect the oxygen-vulnerable polyunsaturated membranes of mitochondria is much greater than previously expected. Thus, an aging cell spends far more of its total energy output as insurance against oxidative damage than a young cell. This concept of shouldering a dual burden of declining energy production by mitochondria and the increased energy cost for protection against oxidative stress define a unified mitochondrial-membrane hypothesis of aging.

1 Mitochondrial Hypothesis of Aging Is Undergoing Revision

The mitochondrial hypothesis of aging states that we age as our mitochondria age. That is, as mitochondria age, their power or energy output is proposed to decrease substantially. Energy is so vital for cells that a program of suicide is activated when energy levels drop below a critical threshold. Recent data using mutator mice genetically engineered for accelerated aging are strongly supportive of the mitochondrial theory of aging (Chapters 8 and 9), but with a new twist. These data show that errors of replication of mitochondrial DNA (mtDNA) generate mutations that accumulate over time, decreasing energy production. Thus, a shrinking supply of energy from mitochondria is proposed to be the root cause of aging. While this pioneering research using mutator mice highlights one important mechanism causing energy deficiency, it does not rule out other possible mechanisms contributing to energy stress. Revision of the conventional mitochondrial theory of aging is now underway and is causing revision of the free radical or oxidative stress theory of aging, the first super theory of aging proposed by Harman and discussed below. The vast surface of mitochondrial and other membranes in human cells contains polyunsaturated fatty acids, which serve as a primary substrate for chemical oxidation (i.e., peroxidation) mediated by free radicals. We suggest that continuous oxidative damage to the membrane contributes directly and indirectly to energy stress, ultimately leading to the demise of the cell. There is no doubt that oxidative stress has powerful effects on our mitochondria, cells, and organisms. Indeed, as described in Chapter 2, the oxidative (free radical) poison paraquat, a cheap farm herbicide, is often chosen for suicide and has been used as a murder weapon. A close look at human death by paraquat shows that this powerful free radical generating poison likely destroys the integrity of energy-transducing membranes of mitochondria. This effect occurs quickly (see Chapter 2), creating catastrophic energy stress, which focuses attention on the famous unanswered "chicken or egg" question of aging and age-related diseases. Which comes first—oxidative stress or energy stress? None of the current theories of aging satisfactorily explain this quandary.

1.1 HISTORICAL PERSPECTIVE

Denham Harman was trained as a chemist at UC Berkeley in the 1950s during the era of free radical chemistry. He recognized the power of free radicals to degrade essential biological molecules such as the highly polyunsaturated membrane building block, docosahexaenoic acid (DHA) (22:6). DHA belongs to the class of oils called omega-3 fatty acids. Harman published his now famous free radical (oxidative stress) hypothesis of aging in 1956, followed sixteen years later with a revised version called the mitochondrial hypothesis of aging, which sets the stage for this book (Harman, 1956, 1972). During this period Harman carried out a series of proof of concept experiments aimed at testing his hypotheses. His research often focused on polyunsaturated fatty acids, including DHA, as possible drivers of the aging process. Indeed, he considered polyunsaturated membranes as being pro-aging in nature. Harman predicted that membrane-soluble antioxidants would reverse the pro-aging effects of DHA and other polyunsaturates. His preliminary data supported this view, but the pro-longevity effects of fat-soluble antioxidants in terms of increasing maximum life span were not substantiated—a disappointment to Harman. However, Harman's publication record shows that linking free radicals, polyunsaturated membranes, and aging was a lifelong interest. He applied his experience with polyunsaturated membranes and their protection in rats to focus on premature aging or degeneration of the human brain. He knew this organ to be enriched with DHA and likely sensitive to free radicals. Harman's later research focused on Alzheimer's disease.

It is clear that Harman considered DHA-enriched membranes to hold an important secret of the brain's aging. It is fair to say that the definitive experimental data or proof of concept for his stimulating hypotheses of aging eluded Harman during his long career. However, historians of science treat Harman as a bright light and founding father of the field of aging.

While Harman was developing his hypotheses and testing the importance of free radicals and polyunsaturated membranes in aging, other pioneers in the field of DHA-enriched membranes were making discoveries seminal toward understanding the benefits and risks of DHA-enriched membranes. For example, in 1967 R. W. Young (see Chapter 12) discovered that specialized membranes called rhodopsin membrane disks found in rod cells of the eye "age" rapidly, requiring replacement on a ten-day cycle. These data were revolutionary at the time because many biologists of this era considered cellular membranes to be chemically stable structures. R. E. Anderson established that DHA is the signature lipid in the specialized rhodopsin disk membranes of the retina, and B. T. Storey showed that DHA in sperm cell membranes is rapidly degraded in the presence of oxygen by a reaction with earmarks of a free radical mechanism (Chapter 12). Storey also found that oxidative damage to DHA inactivated sperm. During this era S. Gudbjarnason reported that DHA levels in heart tissue correctly predicted life span in mammals, an inverse relationship (discussed below and in Chapter 6). We recognize these researchers here for their pioneering studies defining the extraordinary properties of specialized DHA-enriched membranes. Data generated by other pioneers who developed the modern membrane unsaturation theory of aging are highlighted and cited in individual chapters.

1.2 CONVENTIONAL MITOCHONDRIAL THEORY OF AGING IS UNDERGOING REVISION

For decades the mitochondrial theory of aging and oxidative or free radical theories were "joined at the hip," both guiding the field of aging. These two super theories of aging have common elements and share some of the same deficiencies. As the name implies, mitochondria make up the centerpiece for both the conventional and revised mitochondrial theories of aging. According to the conventional mitochondrial theory of aging, oxidative stress was believed to be linked to energy stress through the direct mutagenesis of mtDNA by reactive oxygen species (ROS). This idea has been embraced for many years, supporting a tight and direct linkage between the mitochondrial and oxidative stress theories of aging. The oxidative stress theory has always been the broader of the two in that ROS is proposed to directly target and damage not only DNA, but proteins and membrane lipids as well. ROS-mediated damage to cellular constituents is a chemical reaction well documented in the literature. Data discussed in Section III do not support the view that the major role of ROS in aging is to chemically mutagenize mtDNA. Instead, the revised mitochondrial theory of aging states that errors of replication of mtDNA, not ROS, cause aging (Edgar et al., 2009). These data reinforce the importance of mitochondria or, more specifically, bioenergetically dysfunctional mitochondria as drivers of the aging process.

The severing of the direct linkage between ROS and mutagenesis of mtDNA opens Pandora's box in the field of aging and at first glance seems to diminish the role of oxidative stress in aging. But a closer look gives rise to provocative questions on the possible roles of ROS in aging, as discussed in Section III. If mtDNA is not the primary target of cellular ROS damage during aging, which cellular molecule or structure is the most likely or most important cellular target of ROS in the cell? Since both proteins and membranes are well known as targets of oxidative damage in the cell, the importance of both membranes and proteins is elevated regarding their roles as potential drivers of the aging process.

In essence, data from mutator mice (Figure 1.1) and natural mutators in humans (Chapters 8 and 9) cleave the combined mitochondrial-oxidative stress theory of aging, creating two separate super theories of aging. These data leave proponents of the oxidative stress or free radical theory of aging to ruminate about what molecular roles oxidative stress plays in aging, a point covered in Chapters 11 to 13. These revelations require scientists working in these fields to revisit the mechanisms linking ROS → oxidative stress → aging. A large amount of data is consistent with the oxidative stress theory of aging (Murphy et al., 2011), and it seems likely that oxidative stress plays an important role in aging, as discussed in this book. We suggest that the mtDNA mutator theory of aging, while being revolutionary in scope, should be considered as yet another stepping-stone toward a fuller picture of the sophisticated and multiple cascades governing aging.

1.3   CHANGES HAVE COME TO THE FAMOUS OXIDATIVE STRESS OR FREE RADICAL THEORY OF AGING

For decades the oxidative stress theory of aging, based on the harmful effects of ROS directly targeting and destroying vital cellular components, including DNA, has gained momentum among scientists as a guiding principle or framework for discussion of and planning research strategies to understand aging. Indeed, a monumental quantity of published research has appeared on this subject. Even a few short

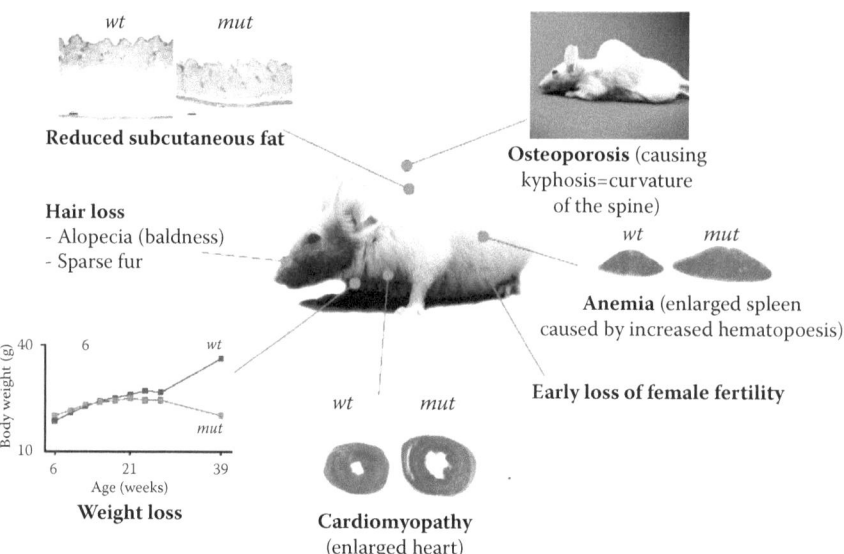

FIGURE 1.1   See color insert. Aging phenotypes in mtDNA mutator mice. The mtDNA mutator mouse is genetically engineered to express a proofreading-deficient version of the mtDNA polymerase. This "mutator gene" leads to a three- to five-fold increase in somatic point mutations of mitochondrial DNA, an occurrence of a linear deleted mtDNA molecule, a progressive respiratory chain dysfunction, an expression of a variety of premature aging phenotypes, and a shortened life span. Humans harboring a naturally occurring mutator gene display numerous pathological phenotypes, as discussed in Chapters 8 and 9. (Reprinted from Edgar et al., *Cell Metab.* 10:131–38, 2009. Copyright © 2009. With permission from Elsevier.)

years ago it would have been difficult to predict that strong winds of change would soon buffet this widely quoted concept. It was a great surprise when data were generated in the early 2000s (Trifunovic et al., 2004) that suggested that the cardinal linkage in the oxidative stress theory of aging might be wrong (Figure 1.1). The heart of the conventional oxidative stress theory can be written like an equation beginning with ROS: ROS → mutations in mtDNA →→→ aging. As introduced above, recent data suggest that ROS do not act directly as mutagens for mtDNA as predicted by the conventional ROS model of aging. Not only has the chain linking ROS and mtDNA mutations been broken, but a new mechanism has taken its place—mainly, that errors of mtDNA replication → mutations of mtDNA →→→ aging. This and other data (Chen et al., 2010) will be covered later, and the main point here is that winds of change are stimulating new ways of thinking about the linkage between oxidative stress and aging, a healthy sign for this field.

1.4   MEMBRANE POLYUNSATURATION THEORY SEEMS TO LINK MITOCHONDRIAL-OXIDATIVE THEORIES OF AGING

Many years ago a Swedish cardiologist arranged to pick up a sample of whale heart for analysis, putting the finishing touches on a new concept he was developing concerning the linkage between the omega-3 fatty acid DHA and aging. Gudbjarnason and colleagues (1978) found that they could predict the heart rate of various mammals based on the DHA content of heart cells. It was known at the time that a tiny mammal such as a miniature vole had a heart rate of greater than 1000 beats per minute, compared to a great whale at 10 to 15 beats per minute. A mechanism to explain why a vole or mouse heart needed so much more DHA in membranes of its cardiac muscle cells, compared to a whale, was unknown at the time. Gudbjarnason reasoned that the answer would be found in the structure-function of the DHA fatty acid chain present in membranes of cardiac muscle cells, and he proposed a novel helical structure for DHA to explain its role.

It was also known at the time that animal size often predicted life span, an inverse relationship. As discussed in Chapter 6, there is evidence that great whales with the lowest levels of DHA in heart cells may be the longest-lived mammals, while the tiniest rodents with much higher levels of DHA have the shortest life spans. This data received relatively little attention until A. J. Hulbert and R. Pamplona, working independently, started their pioneering work leading to the development of the membrane/pacemaker/polyunsaturation/peroxidation theory of aging (Hulbert et al., 2007). This theory states that membrane unsaturation levels in membranes of mammals predict their life span, an inverse relationship (Hulbert at al., 2007). These researchers further proposed that correctly predicting life span from membrane unsaturation levels was based on the chemical properties of the polyunsaturated fatty acid chains present in membrane lipids. The chemical process is called fatty acid peroxidation and occurs whenever unsaturated membranes come into contact with oxygen.

Unsaturated membrane phospholipids, especially DHA or polyunsaturated molecules (the latter found in human mitochondria), are ready targets of peroxidation. Membrane peroxidation is a chemical reaction in which the double bonds of unsaturated fatty acid chains in membrane phospholipids are attacked by molecular O_2. Peroxidation is a free radical-mediated process yielding oxidatively damaged and dysfunctional membranes along with toxic by-products, while increasing the size of the cellular ROS pool. Among the hierarchy of unsaturated fatty acids found in membranes of human cells, DHA is the most sensitive, as follows (in decreasing order of peroxidation rates): DHA (22:6) > EPA (20:5) > ARA (20:4) > linolenic (18:3) > linoleic (18:2) >>> oleic acid (18:1). The numbers in the parentheses refer to the chain length and the number of double bonds per fatty acid chain.

The chemistry of fatty acid peroxidation giving rise to the well-known rancidity of oily foodstuffs also applies to cellular membranes. It is clear that whenever unsaturated fatty acids in human cells come into contact with oxygen, peroxidation will occur and must be contained because of the threat of a catastrophic chain reaction. Free radicals are intermediates in membrane peroxidation chemistry and can build up, reaching levels that cause runaway oxidative reactions, destroy cellular constituents and cells, and even kill humans in some cases (see Chapter 2).

Hulbert and coworkers (2007) considered membrane polyunsaturation theory to be a derivative of the conventional oxidative theory conjoined with the mitochondrial theory of aging. Data showing the severing of the key reaction linking ROS to mtDNA mutations introduced above have left membrane polyunsaturation theory adrift, but not for long. We have recently proposed that the membrane pacemaker theory essentially bridges or joins together the two super theories of aging, helping explain neurodegeneration with implications for aging (Valentine and Valentine, 2013). This unified concept is called the dual-energy pacemaker theory of aging and was initially applied to explain the synergistic roles of oxidative stress and energy stress during neurodegeneration. We now apply the dual-energy pacemaker concept to explain how oxidative stress helps to set the pace of aging. Thus, winds of change have carried in a new view of aging, information that helps unify the understanding of mechanisms underlying human aging and longevity. Note that mitochondria and their specialized polyunsaturated membranes as originally envisioned by Harman remain at the heart of biochemical theories of aging. For a stimulating molecular model linking unsaturated membranes $\rightarrow$ free radicals $\rightarrow$ oxidatively truncated phospholipids $\rightarrow\rightarrow$ mitochondria $\rightarrow\rightarrow\rightarrow$ apoptosis, see McIntyre (2012).

1.5 SUMMARY

An important scientific hurdle to understanding the molecular basis of human aging is to develop a unifying framework to guide research efforts. Ideally, a unified concept should result in more cross-fertilization with other age-related fields, including neurodegeneration and cancer. Currently, there are brilliant islands of knowledge, but it has proven surprisingly difficult to link these realms of knowledge and form a cohesive concept of aging. The field of aging remains plagued by the chicken or the egg problem—involving oxidative stress versus energy stress. Current data suggest that energy stress mediated by mitochondria is the mother of all cellular stresses for aging in mammalian cells, with oxidative and other stresses contributing significantly to energy stress. This quandary is by no means settled and represents a major hurdle in developing a unified concept of aging.

We became interested in the bioenergetics of human aging and longevity following our development in bacteria of the DHA principle, along with a tripartite membrane fatty acid blending code to explain the biochemical roles of DHA and other polyunsaturated chains (Chapter 3). According to the tripartite fatty acid code, human cells incorporate DHA and polyunsaturated fatty acids into their membranes for beneficial purposes, including maximizing membrane motion and energy efficiency. These benefits are balanced against defects generated by conformational dysfunction and oxidative damage inherent in the chemical structures of DHA or polyunsaturated fatty acids, the latter enriched in human mitochondrial membranes. This delicate balance between benefits and risks is called the DHA principle and places all human cells and cells in general into a dilemma. It wasn't long before it became apparent to us that this membrane code had the power to correctly predict why in specific ecological niches one prokaryotic form of life (e.g., Archaea) dominated another (e.g., Bacteria) or vice versa (Valentine, 2007).

The predictive reach of the tripartite membrane fatty acid blending code was next expanded to cover not only prokaryotes but eukaryotes as well (Valentine and Valentine, 2009). After a few years we applied the tripartite membrane fatty acid blending code to neurodegenerative diseases, concluding that Alzheimer's has earmarks of a membrane disease (Valentine and Valentine, 2013). It is well known that the major risk factor for Alzheimer's is aging, and this fact provided a stepping-stone for us in considering the importance of membrane-mediated energy stress on aging (this volume). During this quest we depended heavily on the unity principle of biochemistry to get us across rough terrain where little data are available. According to this concept, the biochemical functions of DHA and other polyunsaturated fatty acid chains working in membranes across life-forms as diverse as deep-sea bacteria and humans are predicted to be similar (Chapter 3). So far we have been encouraged by the predictive powers of the DHA principle and the tripartite membrane fatty acid blending code, and we next apply this reasoning to aging.

REFERENCES

Chen, H., M. Vermulst, Y. E. Wang, et al. 2010. Mitochondrial fusion is required for mtDNA stability in skeletal muscle and tolerance of mtDNA mutations. *Cell* 141:280–89.

Edgar, D., I. Shabalina, Y. Camara, A. Wredenberg, M. A. Calvaruso, et al. 2009. Random point mutations with major effects on protein-coding genes are the driving force behind premature aging in mtDNA mutator mice. *Cell Metab.* 10:131–38.

Gudbjarnason, S., B. Doell, G. Oskardottir, and J. Hallgrimsson. 1978. Modification of cardiac phospholipids and catecholamine stress tolerance. In C. deDuve and O. Hayaishi (eds.), *Tocopherol, oxygen and biomembranes*. Amsterdam: Elsevier, pp. 297–310.

Harman, D. 1956. Aging: a theory based on free radical and radiation chemistry. *J. Gerontol.* 11:298–300.

Harman, D. 1972. The biologic clock: the mitochondria? *J. Am. Geriatr. Soc.* 20:145–47.

Hulbert, A. J., R. Pamplona, R. Buffenstein, and W. A. Buttemer. 2007. Life and death: metabolic rate, membrane composition, and life span of animals. *Physiol. Rev.* 87:1175–213.

McIntyre, T. M. 2012. Bioactive oxidatively truncated phospholipids in inflammation and apoptosis: formation, targets, and inactivation. *Biochim. Biophys. Acta* 1818:2456–64.

Murphy, M. P., A. Holmgren, N. G. Larsson, et al. 2011. Unraveling the biological roles of reactive oxygen species. *Cell Metab.* 13:361–66.

Trifunovic, A., A. Wredenberg, M. Falkenberg, J. N. Spelbrink, A. T. Rovio, et al. 2004. Premature ageing in mice expressing defective mitochondrial DNA polymerase. *Nature* 429:417–23.

Valentine, D. L. 2007. Adaptations to energy stress dictate the ecology and evolution of the Archaea. *Nat. Rev. Microbiol.* 5:316–23.

Valentine, R. C., and D. L. Valentine. 2009. *Omega-3 fatty acids and the DHA principle*. Boca Raton, FL: Taylor and Francis Group.

Valentine, R. C., and D. L. Valentine. 2013. *Neurons and the DHA principle*. Boca Raton, FL: Taylor and Francis Group.

2 Oxidative Stress Defined as a Deadly Free Radical-Mediated Chain Reaction

Case History of Paraquat

Ever since Harmon published his provocative hypotheses of aging (Harman, 1956, 1972), oxidative stress has been considered a cause of aging. In this chapter we borrow a lesson from weed science to explain the nature of and, in this case, the lethal power of an oxidative chain reaction. Paraquat is widely used in the field of aging as a powerful oxidative poison causing a chain reaction of destruction and death in organisms ranging from weeds and *Caenorhabditis elegans* to humans. Paraquat, used by farmers as a broad-spectrum herbicide, has the dubious distinction of being the method of choice for suicide in some countries. This weed killer is inexpensive, and 10 ml of the concentrate can be lethal to humans. Susceptibility to paraquat has been linked to membrane unsaturation levels (Section II). Paraquat is a powerful oxidative toxin in plants and animals, acting as a free radical and short-circuiting high-energy electrons generated during cellular metabolism. The net effect is generation of scorching amounts of reactive oxygen species (ROS), which destroy essential cellular components. Electrons pass from paraquat to oxygen via a chemical reaction and yield one of the most potent chemical toxins found in cells—superoxide radical, in which the oxidative power of O_2 is greatly enhanced after acquiring an additional electron. Superoxide radical is a free radical, and thus paraquat catalyzes formation of free radicals using oxygen as substrate. Superoxide forms even more toxic derivatives, which can chemically attack most cellular constituents, including membranes, proteins, and DNA. Superoxide is a member of a class of naturally occurring reactive oxygen species or free radicals. According to Harman's hypothesis (Harman, 1956, 1972), the continuous destruction of critical cellular components in human cells by free radicals is a cause of aging and age-related diseases. With acute paraquat poisoning it is suggested that a chain reaction of oxidative stress overpowers defenses against oxidative damage, causing rapid-fire cellular death that can quickly spread to destroy functions of vital organs and cause death.

2.1 PARAQUAT GENERATES FREE RADICALS THAT TARGET AND DESTROY CHLOROPLASTS ALONG WITH THE PLANT'S BASIC ENERGY SUPPLY

Rapid death of plants follows spraying with paraquat (Figure 2.1), which triggers a lethal oxidative chain reaction mediated by free radicals. Plants sprayed in the dark remain a healthy green until exposed to light, when large numbers of ROS are produced, causing bleaching of the leaf surface. In Figure 2.1a tiny droplets of herbicide sprayed on a field of weeds drifted on the wind to land on leaves of a nearby cornfield. Thus, chloroplasts, which power plant growth by generation of high-energy electrons, trigger cellular death by passing electrons single file to paraquat, which donates them to oxygen, generating reactive oxygen species (Figure 2.1b). This sparks an oxidative chain reaction that quickly spreads, robs the cells of energy, and shortly kills all of the cells forming the leaf surface. The leaf surface bleached by paraquat is devoid of active chloroplasts, and the photosynthetic energy production necessary for plant growth ceases. We suggest that plant death is caused in part by massive oxidatively mediated destruction of polyunsaturated chloroplast membranes resulting in uncoupling of the chloroplast's bioenergetics.

The roles of high-energy electrons in human diseases are discussed later, but further insight can be gained from looking more closely at paraquat's mode of action in plants. As background, high-energy electrons are generated at photocenters housed in photosynthetic membranes and are passed along the electron transport chain of chloroplasts; these high-energy electrons represent one of the purest forms of energy available to plant cells. Biochemists readily measure the energy content of electrons carried by electron carriers using an oxidation-reduction scale as a standard. As discussed above, paraquat acts in plants by pirating high-energy electrons generated during photosynthesis. In humans paraquat short-circuits electrons originating from the mitochondrial electron transport chain and other specialized reactions where high-energy electrons are generated.

The chemical structure of paraquat is such that only a single electron is accepted and donated. The addition of one high-energy electron converts paraquat, normally in a colorless, oxidized state, into a brilliant-purple reduced state. When aerated, the reduced form of paraquat instantaneously reverts back to its colorless, oxidized state, showing that it has transferred its high-energy electron to O_2. By making chemical substitutions around the dual-ring system of paraquat, synthetic organic chemists have produced derivatives of paraquat that contain electrons with a range of energy levels, covering the high-energy spectrum of biological electrons. Because of these signature changes in color, the Dutch biochemist Albert Jan Kluyver, as early as 1929, used these redox dyes to calibrate which cells in nature handle or produce electrons of the highest energy state.

Kluyver's research inspired us in our discovery of ferredoxin (Valentine, 1964) and development of the high-energy electron concept (Benemann and Valentine, 1971). High-energy electrons generated by chloroplasts during photosynthesis or by bacteria during nitrogen fixation sit near the top of the energy scale for biologically derived, high-energy electrons and readily reduce paraquat. Human cells also generate high-energy electrons capable of reducing paraquat. But these electrons are not

(a)

FIGURE 2.1   See color insert. Paraquat catalyzes formation of reactive oxygen species (ROS), which can kill plants and humans and cause Parkinson's disease. (a) Photo of a corn leaf with lesions bleached by tiny droplets of paraquat drifting on the wind following spraying of a nearby weed field. (Photo used with permission from Dr. Kevin Bradley at the University of Missouri.) *(continued)*

as energetic as those produced during photosynthesis, so the oxidation triggered by paraquat in humans tends to be slower acting than that in plants.

We have gone to some length to describe the nature of high-energy electrons found in various life-forms as background for understanding a principle of oxidative chemistry underlying the oxidative or free radical hypothesis of aging. In searching for potential major sources of available electrons in mitochondria leading to generation of toxic oxidative radicals, the energy state of electrons donated by mitochondria is an important consideration. Note that the highest energy state of electrons destined for driving respiratory energy production is that of the electrons entering the electron transport chain at the level of the high-energy electron site of complex 1 (Chapter 11).

From a chemical perspective, the lethal chemical activity of paraquat resides in its planar dual ring structure, allowing delocalization of a single high-energy electron, which can be pirated from the mitochondrial electron transport chain. Normally such high-energy electrons are carefully shepherded along electron transport chains where their electronic energy can be harnessed directly and in increments to generate proton electrochemical gradients and produce ATP for cellular growth and survival. Paraquat in humans and plants is classified as a nonspecific, high-energy electron acceptor and donor. It short-circuits the electron transport chain of mitochondria in a cyclic manner, first pirating electrons and then passing them to molecular oxygen,

Photochemical Reduction

(b)

(c)

(d)

FIGURE 2.1 (continued) Paraquat catalyzes formation of reactive oxygen species (ROS), which can kill plants and humans and cause Parkinson's disease. (b) Mode of action of paraquat involves short-circuiting of high-energy electrons from the photosynthetic apparatus (photosystem 1.PS1), generating the reduced or free radical form of paraquat. After donating its electron to oxygen and creating a superoxide radical, paraquat is free to accept another high-energy electron, thereby creating a catalytic cycle. Superoxide radical ($O_2^{\cdot-}$) can target unsaturated membranes, generating a lethal oxidative chain reaction. (Image used with permission from the Plant and Soil Sciences eLibrary at the University of Nebraska's Institute of Agriculture and Natural Resources, http://passel.unl.edu.) (c) Structures of ROS formed by paraquat. Note that hydrogen peroxide is not a free radical but is readily chemically converted in the cell to potent derivatives, which are free radicals. (From Held, P., *An Introduction to Reactive Oxygen Species: Measurement of ROS in Cells*, BioTek Instruments, 2010. Copyright © 2010 BioTek Instruments, Inc. and published as an application guide by BioTek Instruments, Winooski, Vt.) (d) Structures of the weed-killer paraquat and the potent neurotoxin MPP$^+$ are similar, and both are now classified as parkinsonian chemicals. Both toxins are proposed to kill cells by a free radical mechanism. (Reprinted from Dauer and Przedborski, *Neuron* 39:889–909, 2003. Copyright © 2003. With permission from Elsevier.)

producing a witch's brew of toxic reactive oxygen species (see Figure 2.1c). As mentioned above, the reduced form of paraquat, itself a free radical, donates one electron to oxygen, generating the potent superoxide radical, which can form even more toxic molecules capable of attacking critical cellular components. Thus, spraying paraquat on a field of weeds on a sunny day is like throwing a spark on a bone-dry pile of shavings, except that paraquat's radiance of flameless chemicals scorches plant cells. In essence, paraquat in the presence of light catalyzes an oxidative chain reaction, breeding potent oxidative free radicals that degrade membranes and any other molecules in their vicinity. It wasn't long before paraquat and similar molecules became suspected as neurotoxins in humans, as discussed next.

2.2   NONLETHAL DOSES OF PARAQUAT AND RELATED CHEMICALS ACT AS NEUROTOXINS AND SELECTIVELY KILL PARKINSONIAN NEURONS

A few days after injecting a contaminated batch of a mood-altering drug he had brewed in a makeshift chemistry laboratory, a young drug user began to shake uncontrollably. The symptoms, though appearing far more quickly, mirrored those of classic age-dependent Parkinson's disease (PD). This led researchers to suspect that a powerful neurotoxin present as a contaminant in the batch of mood-altering drug was the cause (Figure 2.1d). Chemical detective stories pursued almost simultaneously by chemists on both the West and East coasts led to the discovery of a potent, fast-acting neurotoxin called 1-methyl-4-phenyl-1,2,3,6-tetrahydropyridine (MPTP). MPTP is converted to 1-methyl-4-phenylpyridinium (MPP^+) in the brain, where it targets and kills parkinsonian neurons at a rate several orders of magnitude faster than age-dependent Parkinson's disease. From the onset of the discovery of MPTP in the early 1980s, it seemed clear that this neurotoxin must involve a novel mode of action to explain its roughly 1000-fold-faster killing action of parkinsonian neurons.

Young drug users with Parkinson's symptoms had taken intravenously a contaminated derivative of the readily available over-the-counter drug meperidine (Demerol) (Dauer and Przedborski, 2003). This class of drugs has mood-altering effects. A neurotoxic contaminant, MPTP, turned out to be the toxic compound (Langston et al., 1983). Later biochemical studies showed that MPTP undergoes several transformations in the brain, finally yielding MPP^+. There is a great similarity between symptoms of conventional Parkinson's disease and MPTP-induced brain disease. Specifically, both target a small region of the brain, responsible for producing the neurotransmitter dopamine, which is a critical signaling molecule in neurons controlling muscle action. Death of about 60 to 70 percent of these neurons in the substantia nigra compacta (also called parkinsonian/dopaminergic) region of the brain causes symptoms. High-energy electrons generated as a normal part of metabolism in parkinsonian neurons are passed to MPP^+ that acts as a nonspecific electron carrier. Cycling of high-energy electrons through MPP^+ is believed to uncouple energy production (Chan et al., 1991) and generates damaging levels of reactive oxygen species, including superoxide radicals. The rapid drop in energy caused by this neurotoxin seems to fast-forward the death of parkinsonian neurons, compressing the time frame from several decades to merely days or weeks. There is remarkable similarity between the structure and functions of MPTP and paraquat; paraquat is now classified as a parkinsonian chemical.

As discussed in Chapter 20, double bonds of unsaturated fatty acids, such as linoleic acid (18:2), undergoing peroxidation are at the heart of destruction of parkinsonian neurons, data that tighten the linkage between ROS membrane unsaturation and neurodegeneration. The finding that parkinsonian neurons in mice are spared from oxidative death by isotope protection of polyunsaturated chains of membranes (Chapter 20) opens a new window for managing this important neurodegenerative disease, with implications for aging, Alzheimer's disease, cancer, and mitochondrial diseases.

2.3   DEATH AND MURDER BY PARAQUAT

In humans, chronic exposure to paraquat may cause pulmonary fibrosis, a stiffening of the lung tissue. Paraquat causes damage to the body when it touches the lining of the mouth, stomach, or intestines. Paraquat may also damage the kidneys, liver, and esophagus. Sickness may follow if paraquat touches a cut on your skin. If paraquat is swallowed in concentrated form, death can rapidly occur. Death may occur from a hole in the esophagus or from acute inflammation of the mediastinum, the area that surrounds the major blood vessels and airways in the middle of the chest. Sick patients may require a procedure called hemoperfusion, which filters blood through charcoal to try to remove paraquat from the lungs.

In a 1997 paper Stephens and Moormeister (1997) describe four cases of homicidal poisoning by paraquat, three leading to convictions. Oral ingestion is the most common pathway of poisoning, although intravenous paraquat poisoning has been reported and is strongly associated with attempted suicide (Chen et al., 2009). We next explore the molecular basis of acute paraquat poisoning.

Obviously death by paraquat in humans depends on neither light nor chloroplasts (as seen with plants), but does depend on a source of high-energy electrons being generated and transferred in human cells. Since mitochondria routinely receive and handle high-energy electrons, this energy-transducing organelle is considered to replace chloroplasts as the likely active site for paraquat poisoning. In humans a diverse set of organs are damaged during acute paraquat poisoning, although some organs seem to be more sensitive than others. Indeed, specific cells within an organ may be more or less sensitive to cellular death triggered by paraquat, as seen with the case of parkinsonian neurons. Numerous factors, ranging from rates of paraquat uptake and O_2 levels to the levels of Fenton chemistry catalysts, including iron and copper, might modulate patterns of paraquat sensitivity. Mitochondria are surrounded by two different lipid-based membranes called the outer (facing the cytoplasm of the host cell) and the inner (facing the interior or matrix of the mitochondrion) membranes. The inner membrane, which is highly polyunsaturated, represents the clearest target in the cell for membrane peroxidation. We suggest that mitochondrial membrane unsaturation levels are a missing link in understanding paraquat poisoning, as demonstrated by data from model organisms (Chapter 20).

2.4   SUMMARY

Paraquat acts as a rogue electron carrier able to pirate high-energy electrons from the respiratory electron transport chain of mitochondria and pass them singly to O_2, forming a powerful reactive oxygen species called superoxide radical, a free radical. Superoxide radical and its by-products are so chemically reactive that they can trigger an oxidative chain reaction. A ubiquitous set of enzymes has evolved to detoxify superoxide radical, yielding hydrogen peroxide (H_2O_2), a less toxic or reactive form of ROS. In the case of paraquat acting as an herbicide, the killing effect is spectacular, a matter of hours on a sunny day. Paraquat acting in the dark environment of the brain where electrons are significantly less energetic than those of plants spares most neurons, with the exception of neurons of the parkinsonian region of the brain. Parkinsonian neurons seem predisposed toward oxidative stress and killing. After several days the paraquat-like molecule MPTP triggers the killing of a sufficient number of parkinsonian neurons to cause tremors typical of Parkinson's disease. The human body has no defense against high doses of paraquat or MPTP, presumably because of generation of full-blown oxidative chain reactions simultaneously in many critical body organs. Finally, we have chosen to introduce the subject of oxidative stress and the importance of ROS using the case history of acute paraquat poisoning of humans because it highlights the powerful and lethal wave of damage inflicted by a chain reaction mediated by reactive oxygen radicals. The take-home lesson is that maintaining ROS in the human body at nontoxic levels is an essential strategy for preventing an oxidative chain reaction that can destroy mitochondria, their host cells, and eventually the brain or whole body.

REFERENCES

Benemann, J. R., and R. C. Valentine. 1971. High-energy electrons in bacteria. *Adv. Microb. Physiol.* 5:135–72.

Chan, P., L. E. DeLanney, I. Irwin, et al. 1991. Rapid ATP loss caused by 1-methyl-4-phenyl-1,2,3,6-tetrahydropyridine in mouse brain. *J. Neurochem.* 57:348–51.

Chen, H. W., T. K. Tseng, and L. W. Ding. 2009. Intravenous paraquat poisoning. *J. Chin. Med. Assoc.* 72:547–50.

Dauer, W., and S. Przedborski. 2003. Parkinson's disease: mechanisms and models. *Neuron* 39:889–909.

Harman, D. 1956. Aging: a theory based on free radical and radiation chemistry. *J. Gerontol.* 11:298–300.

Harman, D. 1972. The biologic clock: the mitochondria? *J. Am. Geriatr. Soc.* 20:145–47.

Langston, J. W., P. Ballard, J. W. Tetrud, et al. 1983. Chronic parkinsonism in humans due to a product of meperidine-analog synthesis. *Science* 219:979–80.

Stephens, B. G., and S. K. Moormeister. 1997. Homicidal poisoning by paraquat. *Am. J. Forensic Med. Pathol.* 18:33–39.

Valentine, R. C. 1964. Bacterial ferredoxin. *Bacteriol. Rev.* 28:497–517.

3 Membranes of Deep-Sea Bacteria as Surrogates for Mitochondrial Membranes of Humans

Bacteria first became prominent as model organisms for studies of fundamental aspects of aging and age-related diseases during the last several decades of the twentieth century. During that era data from bacteria seemed to reinforce the popular (conventional) mitochondrial theory of aging featuring oxidative stress acting as a mutagen and directly driving the aging process. For example, data from bacteria showed that reactive oxygen species (ROS) could chemically modify bacterial DNA and cause mutations—a finding consistent with a direct linkage between ROS, mutations, and aging.

The favorite bacterial species for research on fundamental aspects of aging and cancer were *Escherichia coli* and *Salmonella*, whose membrane fatty acid building blocks (phospholipids) are monounsaturated in contrast to polyunsaturated membranes of human mitochondria. Pioneering researchers at the time showed that monounsaturated membranes of *E. coli* and *Salmonella* contribute to the pool of ROS and apparently act directly as chemical mutagens. Human mitochondrial membranes are estimated to be 30 to 40 times more sensitive to oxidative damage (peroxidation) than membranes of *E. coli*. Cardiolipin (CL), a highly polyunsaturated molecular species of phospholipid in mitochondria, might be 50-fold more sensitive to peroxidation than monounsaturated CL of *E. coli*. Furthermore, *E. coli* and many other bacteria have evolved a mechanism involving cyclopropane fatty acid formation, which protects their monounsaturated chains against peroxidation, a mechanism absent in human mitochondria.

We have observed that natural isolates of *E. coli* obtained from hospital collections can derivatize >90 percent of their total monounsaturated fatty acid chains in membranes with cyclopropane protective groups. Other bacteria produce membranes even more oxidatively stable than *E. coli* by eliminating all unsaturation from their membranes and replacing unsaturated chains with methyl-branched chains, which, like cyclopropane fatty acids, are orders of magnitude more resistant to oxidation. The main point is that bacteria used initially as models for studies of the membrane's contribution to aging have evolved membranes highly resistant to peroxidation, and thus lack properties found in polyunsaturated membranes of human mitochondria.

3.1 OXYGEN-DAMAGED MEMBRANE FATTY ACIDS SCORE POSITIVE AS MUTAGENS IN THE AMES CARCINOGEN TEST

Bruce Ames and colleagues developed *Salmonella* tester strains in the 1970s for detection of mutagenic chemicals present in the environment (Figure 3.1). This assay has proven to be ultra-sensitive for this purpose. Various reactive oxygen species (ROS), including fatty acid peroxides, by-products of membrane oxidation, score positive using the Ames mutagen test. At the time, these data strongly reinforced the concept that ROS produced by mitochondria directly mutagenize mitochondrial DNA (mtDNA). These data also clearly showed that ROS had the potential to act as a mutagen for chromosomal DNA, perhaps causing cancer. After three decades the direct role of ROS → mtDNA as the cause of aging remains controversial.

Convincing data using mutator mice, since reproduced in other laboratories (see Chapter 9), show that errors of mtDNA replication, not ROS, cause mutations in mtDNA. According to this revised model, mutations accumulate in a linear fashion in mtDNA during aging, eventually decreasing energy output by mitochondria to a critical threshold triggering massive programmed cellular death. The data derived from mutator mice severed the former mitochondrial theory of aging into two separate theories, one featuring energy stress and the other oxidative stress. These data open Pandora's box concerning the order and roles played by energy stress versus oxidative stress in aging.

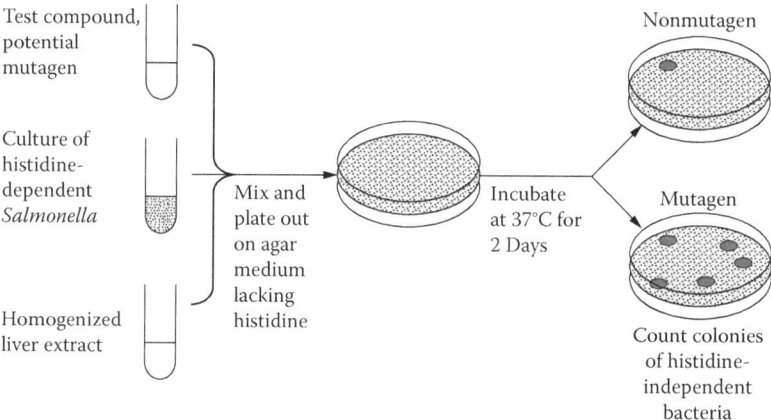

FIGURE 3.1 Oxidation products of unsaturated fatty acids score positive in the Ames test for mutagenicity. The test uses a strain of *Salmonella* bacteria that requires histidine in the medium because of a defect in a gene necessary for histidine biosynthesis. Mutagens present in the environment reverse the original genetic defect, creating revertant bacteria that do not require histidine. Data from this highly sensitive test were consistent with the popular hypothesis at the time that ROS act directly to mutate mitochondrial DNA and cause aging. This concept is now being reevaluated. (From Alberts et al., *Molecular Biology of the Cell*, 5th ed., 2007. Copyright © 2007. Republished with permission of Garland Science-Books; permission conveyed through Copyright Clearance Center, Inc.)

In attempting to sort out the contributions of bacteria in the field of aging, we go back to data generated by the Ames test itself. Is it possible that the *Salmonella* tester strain, being ultra-sensitive in detecting mutagens, led researchers to a false conclusion in the case of ROS acting as a mutagen and directly targeting mtDNA? There is another interesting point regarding the history of bacteria as model organisms for understanding aging that illuminates this question.

As mentioned above, the membranes of the bacteria used as models for studying aging during the twentieth century are dramatically different in fatty acid unsaturation levels and oxidative stability than in mitochondrial and other highly unsaturated membranes of a mouse or a human. For example, light sensing rhodopsin membrane disks in rod cells of the human eye represent an extreme case because their specialized membranes are significantly enriched with di-docosahexaenoic acid, a molecular species of membrane phospholipids with twelve (carbon-to-carbon) double bonds per molecule. In contrast, *E. coli* or *Salmonella*, favorite tools for earlier research on aging, synthesize phospholipids with an average of a single double bond per molecule in cells grown at 37°C and two double bonds in a major class of phospholipids found in cells adapted to grow at 20°C.

Data establishing a linkage between membrane unsaturation and oxidative stress derived by Ames and others using bacteria are still relevant today. But in recent years bacteria have received relatively little attention as suitable model organisms for understanding what is now called the membrane pacemaker, polyunsaturation, or peroxidation theory of aging (Chapter 6). Thanks to the discovery of docosahexaenoic acid (DHA)- and EPA-producing bacteria from the deep sea, we are able to revisit bacteria as the simplest model organisms for understanding the polyunsaturated membrane's contribution to aging (Figure 3.2).

3.2   LESSONS FROM DHA/EPA-PRODUCING BACTERIA

In 1986 several isolates of deep-sea bacteria were discovered to synthesize the prominent omega-3 fatty acid DHA, which was shown to be incorporated into membrane phospholipids (DeLong and Yayanos, 1986). Subsequent genomic analyses of two different isolates of DHA-producing strains, *Moritella* (Yayanos et al., 2007) and *Colwellia* (Methé et al., 2005), provide a picture of the lifestyle of these marine bacteria, along with clues about the benefits and risks of DHA working in membranes of these cells (Valentine and Valentine, 2009). DHA-producing bacteria are adapted to a world characterized by extreme cold and great hydrostatic pressure, along with high salinity of seawater. Not only are these bacteria adapted to these harsh or extreme conditions, but in some cases they have become strictly dependent on deep-sea conditions for growth and die when their normal extreme conditions are significantly altered. For example, isolates of some DHA bacteria from the deep sea grow within a temperature range from 0 to 10°C and are killed by temperatures above 10°C, while others grow to a temperature of 20°C, with death occurring at a few degrees above the maximum.

DHA-producing isolates require seawater levels of salinity for growth, along with cold temperatures. The requirement for seawater levels of Na^+ for growth can be

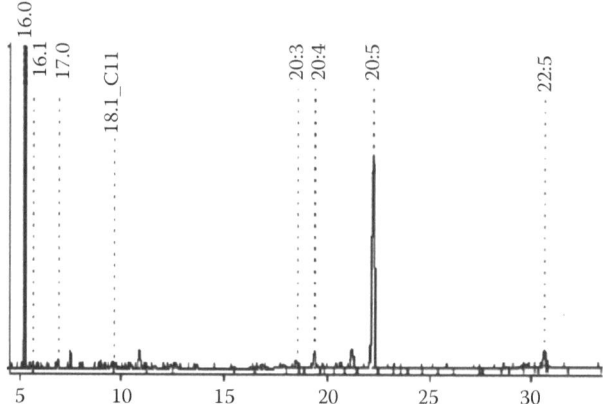

FIGURE 3.2 EPA-enriched membranes of a recombinant of *E. coli* serve as a surrogate for polyunsaturated membranes of mitochondria. This membrane fatty acid profile of an EPA recombinant of *E. coli* shows that EPA replaces virtually all of the 16:1 and 18:1 molecular species of phospholipids, normally making up about 50 percent of total membrane fatty acids in this cell. EPA recombinant cells of *E. coli* synthesizing greater than 30 percent EPA as total fatty acids display unexpected properties, data that led us to propose the DHA principle and the tripartite membrane fatty acid blending code discussed in this chapter. (From Valentine and Valentine, *Omega-3 Fatty Acids and the DHA Principle*. Boca Raton, FL: Taylor and Francis Group, 2009. Copyright © 2009. Reproduced with permission of CRC Press; obtained via Copyright Clearance Center, Inc.)

explained in part by the evolution of Na⁺ bioenergetics in these bacteria (reviewed by Valentine and Valentine, 2009). Specifically a novel form of the powerful proton pump, NADH dehydrogenase (i.e., complex 1 in mitochondria), has evolved as a primary mechanism to efflux Na⁺ out of the cell for the purpose of generating and maintaining an electrochemical gradient of Na⁺ across the membrane (high outside, low inside). This Na⁺ gradient energizes flagellar motion, drives ATP synthesis by ATP synthase, and powers numerous metabolite uptake pumps located in the membrane. A large battery of substrates can donate electrons for conventional respiration or anaerobic respiration, the latter utilizing a number of organic/inorganic molecules as terminal electron acceptors in place of oxygen. Fermentation has been described for some isolates, but respiration seems to be the primary energy-producing mode, with active rates of respiration reported at temperatures approaching 0°C.

DHA biosynthesis in these bacteria occurs by a novel anaerobic mechanism called the polyketide pathway. Numerous secondary metabolites, most notably several different antibiotics, are known to be produced by a polyketide mechanism in various microbes, but this is a first for a fatty acid (Metz et al., 2001). In the case of the DHA chain, elongation and insertion of double bonds in the chain occur in the absence of molecular oxygen, an essential substrate for desaturation of fatty acids by the conventional desaturase enzymes of plants and animals. We suggest that the anaerobic pathway of DHA biosynthesis offers several benefits to DHA bacteria, which often live in low-oxygen or anaerobic environments such as the gastrointestinal tracts of fish or mollusks. These advantages include:

- Enhanced rates of electron transport at cold temperatures in both oxygenated and anoxic environments
- A membrane permeability barrier biased toward Na^+ ions over H^+ and dependent on cold temperatures, high salinity, and often high hydrostatic pressure to maintain membrane architecture suitable to plug cation leaks
- Growth in low-oxygen environments, which stabilizes DHA membranes by avoiding DHA oxidation, and prevents membrane dysfunction

The main conclusion at this point is that DHA offers both benefits and risks to deep-sea bacteria, giving rise to the DHA principle (Valentine and Valentine, 2009), which we propose is applicable to polyunsaturated membranes of mitochondria.

DHA/EPA-producing bacteria do not form endospores, the most hardy, stress-resistant, and longest viable bacterial state. Endospores are thought to persist in a metabolically dormant state. Tolerant states of marine bacteria use a different survival strategy, slowing down but not stopping their metabolism during long periods when conditions are not favorable for rapid vegetative growth. When starved of nutrients vegetatively, growing cells of marine bacteria often convert to a miniaturized stage, able to carry out maintenance metabolism with greatly extended times between division cycles. This mechanism of miniaturization-maintenance metabolism is widespread among marine organisms (obviously not whales, which use gigantism for survival) and allows marine bacteria to persist for at least several years in seawater, which is typically low in organic nutrients. Genomic analysis suggests that marine bacteria have evolved multiple mechanisms to protect their membranes against peroxidation during different stages of their life cycle. These mechanisms include protection of monounsaturated but not polyunsaturated chains by cyclopropane formation, and biosynthesis of methyl-branched chains; both mechanisms lead to fatty acids resistant to peroxidation. Note that cyclopropane synthetase does not utilize either polyunsaturated fatty acids (PUFAs), of two to three double bonds, or highly unsaturated fatty acids (HUFAs), with four to six double bonds per chain, as substrates.

DHA/EPA-producing bacteria have also evolved powerful regulatory systems governing the timing of synthesis of DHA/EPA, another mechanism to maximize benefits of these reactive fatty acids. For example, as temperatures approach a maximum for supporting growth, DHA/EPA levels in membranes begin to plummet, often reaching undetectable levels at maximum growth temperatures. Increasing salinity has the opposite effect, counterbalancing the down-regulation by increased temperature and, in some cases, shifting the maximum growth temperatures to higher levels. Anaerobic growth conditions, including anaerobic respiration, which avoid peroxidation, may also up-regulate levels of DHA/EPA in membranes. These data are consistent with timing of DHA/EPA synthesis toward maximizing benefits while reducing peroxidative or conformational risks from these chains.

The convergent evolution of an anaerobic pathway of DHA/EPA synthesis can be explained in part as a mechanism to avoid peroxidation of these chains. Genomic analysis covering hundreds of different bacteria living in a range of habitats on land and sea shows that DHA genes are present in bacteria growing in specialized niches of the biosphere, primarily the extremely cold, saline, and highly pressurized marine environments. The unique ecological distribution pattern of these bacteria can be explained in part by their need to avoid environments supporting high rates of membrane peroxidation and favoring habitats in which membrane peroxidation is minimized. We suggest that all organisms producing DHA or EPA face a delicate balancing act between benefits and risks. Again, these properties help define the DHA principle.

3.3 TRIPARTITE MEMBRANE FATTY ACID BLENDING CODE CAN EXPLAIN BIZARRE PROPERTIES OF AN EPA RECOMBINANT OF *E. COLI*

Cloning and expression of EPA genes from a marine bacterium in *E. coli* were first reported in the mid-1990s (Yazawa, 1996). The first generation of EPA recombinants of *E. coli* expressed EPA genes, but incorporated into their membranes only a few percent of EPA as total membrane fatty acids. This modest level of EPA did not significantly change the bulk fatty acid profile or properties of EPA recombinant cells, which essentially retained a membrane whose properties are similar to wild-type *E. coli* cells. However, this picture changed dramatically when Jim Metz at Calgene transformed an unsaturated fatty acid-requiring mutant of *E. coli* with a plasmid carrying the EPA gene cluster provided by Yazawa and colleagues (Metz et al., 2001) (Figure 3.2). The recipient used to construct the EPA recombinant was an *E. coli* fabB⁻ mutant that carries a targeted transposon-induced knockout mutation of a gene required for synthesis of monounsaturated fatty acids. This mutation blocks growth in the absence of a dietary source of unsaturated chains.

Normally membranes of *E. coli* grown at 37°C are composed of roughly 50 percent of total fatty acids as long-chain monounsaturates (16:1/18:1). The *E. coli* fabB⁻ mutant retains the ability to synthesize saturated fatty acids of long chain length, but in the absence of unsaturated chains it produces a dysfunctional membrane. The term *dysfunctional* refers in this case to a membrane composition that does not meet the structural and biochemical properties needed to sustain cellular activity. Feeding unsaturated fatty acids with from one to five double bonds, but not six, as in DHA, restores growth. Metz found that expression of EPA genes carried on a plasmid also restores growth of the unsaturated fatty acid-requiring mutant and results in surprisingly high levels of incorporation of EPA into membrane phospholipids (Metz et al., 2001).

Figure 3.2, a fatty acid profile of membranes of the EPA recombinant, shows the presence of >30 percent EPA as total membrane fatty acids and the virtual absence of monounsaturated classes of fatty acids, the latter normally being the major unsaturated species present. The major saturated fatty acid chain is 16:0; significant levels of 14:0 not shown on this chromatograph are also present in EPA recombinant cells. Small peaks for 20:3, 20:4, and 22:5 are present and likely are derived as by-products of EPA synthetase being overexpressed in the EPA recombinant. Note that the beta-oxidation pathway present in wild-type *E. coli* is mutationally blocked in EPA recombinant cells. This simplifies interpretation of the biochemical roles of EPA, which are limited in recombinant cells to structural membrane functions.

The presence of EPA as the bulk unsaturated fatty acid in membranes of the EPA recombinant shows clearly that EPA chains contribute membrane fluidity essential for growth of *E. coli*. Using growth assays, we have calibrated that one chain of EPA in membranes of *E. coli* contributes fluidizing power for growth equivalent to about two monounsaturated chains (Valentine and Valentine, 2009). Preliminary micro-biological and physiological analysis of the EPA recombinant revealed a number of extraordinary properties that might be attributed to replacement of monounsaturated chains with high levels of EPA in membrane phospholipids, as follows:

- EPA is a far more powerful fluidizing molecule than monounsaturated chains and supports active respiration to 4°C and perhaps lower.
- EPA partnering with 16:0 in phospholipids generates membrane molecular architecture supporting respiratory growth and proton bioenergetics. (Note that direct measurement of proton leakage across these membranes or ves-icles derived from these cells has not been conducted, but we suspect that rates of both respiration and proton leakage are elevated in EPA recombi-nant cells.)
- EPA recombinant cells of *E. coli* are unable to grow fermentatively or by anaerobic respiration due to an unknown defect of metabolism.
- EPA recombinant cells growing in deep, agar stabs are initially strongly inhibited by ambient levels of oxygen at the surface and prefer a micro-aerophilic environment found deeper in the stab.
- EPA recombinant cells are strictly cold dependent for growth (i.e., growth range restricted to 13 to 21°C, in contrast to 8 to 45°C for wild-type *E. coli* cells).
- Growth of EPA recombinant cells near their maximum growth temperature but not at cooler temperatures is dependent on seawater levels of salinity, with no growth occurring in classic Luria broth agar minus NaCl. Wild-type cells grow rapidly under these conditions. Optimal growth of recom-binant cells occurs at about 0.35 M NaCl (i.e., roughly seawater levels), with higher levels causing inhibition. Maximum temperature for growth can be expanded from about 21 to 24°C in growth medium supplemented with 0.35 M NaCl.
- Compared to wild-type cells EPA recombinant cells are extremely sensitive to photooxidative killing by singlet oxygen.
- Curing the recombinant of the EPA plasmid restores the original *E. coli* fabB⁻ phenotype, which is genetically stabilized by the presence of the EPA plasmid.

These physiological properties displayed by the EPA recombinant are remarkably different than those of wild-type *E. coli*, and instead closely resemble characteristics of a typical marine bacterium. These data show that enrichment of the membrane by EPA has a powerful influence on the ecology of *E. coli* and likely cells in general. The above exploratory experiments on the EPA recombinant were carried out at Calgene in the mid-1990s as a side project to the main effort of expressing DHA/EPA genes from bacteria in new oil crops. Thus, the novel phenotypes and physiological properties listed above were not traced to their biochemical roots. Yet, as a whole, these data can be explained by the DHA principle. These data led to the development of a tripartite membrane fatty acid blending code (FA code), which we suggest is universal. According to this code, DHA or EPA contributes at least three biochemical properties to membranes—extreme motion, a mediocre permeability barrier against protons with a dependency on and a bias toward Na^+ over protons as energizing ion, and a great sensitivity to oxidative damage.

3.4 THE DHA PRINCIPLE HAS A WIDE RANGE OF APPLICATIONS

The DHA principle is generalized here to cover mitochondrial membranes and states that polyunsaturated membranes are always a result of molecular compromise, offering both benefits and risks to cells. DHA sits at the top of a hierarchy of unsaturated fatty acids in nature and has the most extraordinary properties and dynamic conformations among all fatty acid chains. In spite of their extraordinary physical-chemical properties, we suggest that DHA chains still must fulfill the same fundamental membrane functions as do other classes of membrane fatty acids. We hypothesize that lessons learned on the structure-function of DHA apply not only to EPA, but also to membrane fatty acids in general, including mitochondrial membranes. Thus, the DHA principle is considered a universal rule applicable to membranes of all life-forms.

To test the DHA principle from a global, ecological perspective, we first considered the unique distribution pattern of DHA/EPA in the biosphere. We estimate that annual global photosynthetic production of DHA/EPA by marine phytoplankton is in the range of several hundred million metric tons annually. In contrast, land plants, representing at least half of the annual global primary production, do not produce or incorporate DHA/EPA into their membranes, and we argue that DHA/EPA would be toxic to most terrestrial plants (Valentine and Valentine, 2009). Mosses are an exception, producing both ARA (20:4) and EPA (20:5), but not DHA. Thus, the vast majority of global DHA/EPA is produced by eukaryotic phytoplankton and occurs in cool to cold oceanic ecosystems. In contrast, prokaryotic phytoplankton such as cyanobacteria, which dominate a vast area of tropical and subtropical seas and contribute about 20 percent of total fixed carbon, are devoid of DHA/EPA in their membranes, similar to land plants.

We propose that the DHA principle is at work on a global scale, with cold-adapted marine plants benefiting from DHA/EPA in contrast to land plants, where the risk of DHA/EPA exceeds the benefit. We have identified numerous similar examples among bacteria, plants, and animals in which the benefits versus risks of DHA/EPA appear to dictate the ecological distribution pattern of organisms that produce these fatty acids. For example, coral decline caused by warming ocean temperatures can be explained by a mechanism in which oceanic warming of even a few degrees tips the balance between the benefit versus risk of omega-3 fatty acids enriched in membranes of coral algal symbionts, first killing the symbiont and then the coral (Tchernov et al., 2004). As discussed later, the levels and distribution patterns of DHA in tissues of a short-lived mouse are dramatically different than the distribution pattern in a long-lived human, with most human cells containing only traces of

DHA compared to the high levels of DHA found in cells of mice. We suggest that the DHA principle helps explain why organisms choose specific kinds of fatty acids for their membranes. We next discuss the biochemical rationale behind the DHA principle, with bacteria once again providing valuable lessons.

3.5   TRIPARTITE MEMBRANE FATTY ACID BLENDING CODE IS ALSO LIKELY UNIVERSAL

Data from DHA/EPA bacteria provide a new perspective on the fundamental biochemical or molecular roles played by polyunsaturated fatty acid chains in membranes. Having satisfied ourselves that the DHA principle correctly predicts patterns of DHA/EPA production, ranging from a global to a subcellular scale, we began a quest to decipher the membrane fatty acid code at the molecular or biochemical level. This quest was greatly aided by physiological data accumulated from studies of the bizarre properties of EPA recombinant cells discussed above. From these and other data in the literature, we predict that DHA/EPA contributes at least three fundamental biochemical properties essential for membrane function, all being a reflection of the extraordinary physical/chemical/biochemical properties of these chains, as follows:

- Extreme lateral and rotational motion, which maximizes collisions of membrane components and maintains a functional physical state of the bilayer even under deep-sea conditions
- A mediocre permeability barrier for bioenergetically important cations (i.e., H^+, Na^+, and K^+) with a bias toward Na^+ bioenergetics
- Extreme susceptibility to membrane peroxidation avoided by growth of bacteria in the cold and under anaerobic conditions, otherwise requiring novel protection mechanisms against peroxidation in the presence of air

DHA bacteria have been found to catalyze rates of respiration at 4°C that are about 50 percent the rate recorded at 15°C, their maximum growth temperature. We have similarly found that the rates of respiration by the EPA recombinant at 4°C are about 40 percent the rate, compared to 15°C (Valentine and Valentine, 2009). We have also found that about 12 percent of EPA as total membrane fatty acids is sufficient to support growth of EPA recombinant cells at 24°C. The level of saturated chains in recombinant cells under these conditions is nearly 80 percent, mostly 16:0. Around 25 to 30 percent monounsaturated chains are normally the minimal levels required for supporting growth of *E. coli* at 24°C.

We suggest that the anaerobic pathway of DHA/EPA biosynthesis evolved in marine bacteria as a mechanism to maintain membrane motion, enabling or maximizing respiratory energy production and other membrane processes under extreme conditions encountered in the deep sea. Thus, bioenergetic gain seems to be the selective force behind evolution of a novel convergent, anaerobic pathway of DHA/EPA production with the sea as the proving ground. We further propose, based on studies with bacteria, that DHA/EPA behaves as a powerful membrane antifreeze harnessed to speed up rates of electron transport and energy production in both ectothermic and endothermic organisms. As discussed in later chapters, the absence of DNA in mitochondrial membranes of humans can be explained by the tripartite membrane fatty acid blending code.

There are increasing data that DHA/EPA-enriched membranes are leaky for bioenergetically important cations—H^+, Na^+, and K^+. Since membranes act as gate-keepers for preserving electrochemical gradients of protons and sodium, spontaneous leakage across the lipid portion of the bilayer robs the cell of energy. As already mentioned, cold temperatures and high pressure may conformationally adjust the membrane permeability barrier of DHA/EPA bacteria, preventing futile cation leakage.

A number of other mechanisms have evolved to decrease futile cycling of cations across bacterial membranes, including appending bulky methyl groups on fatty acid chains as described above. High Na+ levels, as mentioned above and described previously (Valentine and Valentine, 2009), modulate synthesis of DHA/EPA and may directly tighten the cation permeability barrier of membranes. In animals, cholesterol is thought to act by plugging cation leakage of membranes (as discussed in Chapter 10). In the early 1980s it was found that protons pass spontaneously across membranes at amazing rates by a novel proton (cation) tunneling mechanism (Nichols and Deamer, 1980). Controversial at first, this mechanism has been confirmed during studies of the mode of action of gramicidin A. This peptidic antibiotic kills target bacterial cells by catastrophic energy uncoupling caused by breaching of electrochemical gradients of protons essential for energizing the cell. Gramicidin A kills by forming a membrane-spanning nanotube with a core region containing a thread of water formed by 21 water molecules arranged in single file and matching the thickness of the bilayer. Each water-wire completes a circuit connecting high proton levels outside the cell with the low proton levels in the cytoplasm. An open circuit permits protons to hop along the water-wire from high to low concentrations at rates of ~20,000 protons each millisecond. Na^+ ions also pass along a water-wire, but rates are three to four orders of magnitude slower than those of protons. Nevertheless, the slower rate of transport of Na^+ is overcome to a large extent by the comparatively much greater concentrations of sodium surrounding many cells.

Water-wires are thought to form spontaneously in all membranes dependent on variables including temperature, unsaturation levels, membrane thickness, hydrostatic pressure, salinity levels, and the presence of damaged or dysfunctional lipid structures. Whereas water-wires have been clearly defined in the case of the mode of action of gramicidin A, their existence in natural membranes is based on convincing but indirect evidence. For example, the relative rates of cation passage crossing membranes ($H^+ >>> Na^+ > K^+$) mirror that of the gramicidin A system, consistent with a water-wire mechanism. As mentioned above, Na^+ movement via water-wires is several orders of magnitude slower than that of protons. This large difference in permeability between protons and sodium is leveraged for bioenergetic gain by bacteria living in niches where proton bioenergetics is chronically compromised. Numerous bacteria have switched to Na^+ bioenergetics, presumably to avoid chronic proton uncoupling of their membranes. A selected list of prokaryotic cells using Na^+ bioenergetics follows:

- A heat-adapted anaerobic bacterium living in a hot spring near the maximum temperature supporting bacterial life
- A petroleum-degrading marine bacterium growing in intimate contact with toxic oil droplets containing toluene and other membrane-destroying solvents
- DHA/EPA-producing bacteria living under energy stress in marine environments
- Methanogenic and other Archaea

The common stress shared by these cells is energy stress caused in part by futile proton cycling induced by the environment. We suggest that by switching to Na^+ bioenergetics, bacteria conserve significant amounts of energy at the membrane level, enabling their growth in niches in which excessive proton leakage is unavoidable. Bacteria appear to have evolved at least four different mechanisms to protect against excessive proton leakage: change membrane architecture (e.g., add bulk as extra methyl groups), produce more energy to counterbalance leaks, modify former proton-specific sites of respiratory and other cation pumps to accommodate and efflux Na^+, and evolve new primary Na^+ pumps such as Na^+-translocating NADH dehydrogenase specific for Na^+ or Li^+. Among prokaryotic organisms, Archaea produce the most robust membranes against proton/sodium leakage and exploit this advantage to grow in extreme environments where they often outcompete bacteria (Valentine, 2007). Note that Archaea carry out both proton and sodium bioenergetics. Thus, genomic, ecological, physiological, and biochemical data from Bacteria and Archaea are consistent with the concept that water-wires spontaneously forming in membranes and dependent on lipid dynamics and chemistry play fundamental roles in bioenergetics of these cells. We suggest that lessons on the contributions of membrane lipids to bioenergetics of prokaryotic cells are applicable to energy-transducing organelles, especially chloroplasts and mitochondria, as discussed next.

3.6   APPLICATIONS TO MITOCHONDRIA

Mitochondria retain many features of bacteria but have left behind many of the mechanisms used by bacteria to protect their membranes against oxidative stress and energy uncoupling. Once again, some of the membrane-based mechanisms commonly used by bacteria to prevent leakage and conserve their H^+/Na^+ electrochemical gradients are as follows:

- Modify membrane fatty acid composition or acyl head group linkages
- Inhabit environmental niches in which membrane molecular architecture is dictated by and dependent upon environmental parameters, especially low temperature and high salinity
- Switch to Na^+ bioenergetics
- Inhabit ecological niches where energy supply is usually in excess, allowing these cells to produce more energy than is wasted by futile energy cycling across membranes

The first three of these mechanisms have already been discussed, and the attention now shifts to the fourth point. This strategy, dependent on an abundant source of energy present in the environment, has allowed some bacteria to maintain a constant or unchanging membrane fatty acid composition while growing over a range of temperatures where proton leakage is known to increase as a function of temperature (van de Vossenberg et al., 1998, 1999). *B. stearothermophilus*, a moderate thermophile, synthesizes a compositionally invariant membrane and uses a novel but energy-intensive mechanism to survive at relatively high temperatures around 50°C (De Vrij et al., 1988). In spite of major proton uncoupling caused by conformational dysfunction of the membrane permeability barrier at high temperature (van de Vossenberg et al., 1999), *B. stearothermophilus* increases its respiration rate as temperatures rise as a means to increase energy output. Thus, energy output stays ahead of futile proton cycling. Once again, this strategy, being energy wasteful, depends on a continuous and plentiful supply of energy in the environment. In contrast, most bacteria adapt to significant changes in temperature by altering membrane composition, a mechanism that saves a great deal of energy (van de Vossenberg et al., 1999), and thus presumably provides a competitive advantage.

We suggest that mammalian mitochondria have evolved and refined a bioenergetic strategy along the lines of *B. stearothermophilus*. We hypothesize that an increased demand for energy (energy stress) in mitochondria is equivalent to energy stress caused by thermally mediated energy uncoupling in *B. stearothermophilus*. We further suggest that an invariant membrane composition is a selective advantage to human mitochondria because it eliminates the need and associated costs of constantly changing membrane lipid composition to keep up with a rapidly changing and often high-energy demand. The similarities between bioenergetic strategies of *B. stearothermophilus* and mitochondria can be summarized as follows:

- Invariant membrane fatty acid composition.
- A requirement in mitochondria for a consistent and abundant supply of energy provided by the host cell.
- In *B. stearothermophilus*, high temperature increases rates of respiration, energy production, and proton leakage, whereas human mitochondria operating at a constant 37°C have evolved to respond quickly to signals reflecting changes in energy demand by modulating rates of respiration. Mitochondria, especially in ectothermic animals, are also capable of membrane adaptation in response to changes in the environment and energy demand. These adaptations often involve modulation of levels of membrane unsaturation, including enrichment with DHA/EPA.

It is clear that Darwinian selection occurring over almost a billion years has honed mitochondrial energy efficiency to its present remarkable level. It is also clear that such a high level of mitochondrial energy efficiency is dependent on the many advantages of an intracellular location where idealized homeostatic conditions are maintained by the host cell and host organism. The list of advantages includes homeostasis for temperature, pH, oxygen, ion composition, osmolality, energy substrates, and myriad other advantages not shared with free-living bacteria. These properties likely endow the inner membrane of mitochondria with a more robust or efficient permeability barrier against futile cycling of protons via a water-wire mechanism (Jastroch et al., 2010; see Chapter 10). In spite of their "luxurious" surroundings inside cells, mitochondria are often required to work at intense rates, and their membranes can degrade rapidly, requiring constant renewal. We suggest that aging of human mitochondria is caused in part by the continuous degradation of their extraordinary polyunsaturated membranes, which share properties with energy-transducing membranes of DHA/EPA-producing bacteria.

3.7 BACTERIAL RECOMBINANTS AND MUTANTS DESERVE MORE ATTENTION AS MODELS FOR MITOCHONDRIAL MEMBRANES

The final point in this chapter focuses on the future role of bacteria as model organisms for studying aging. As highlighted in a previous book (Valentine and Valentine, 2009), we suggest that DHA/EPA working in membranes plays the same fundamental roles assigned to membrane fatty acids in general—permeability, motion, and stability. The caveat is that DHA/EPA is specialized or extraordinary in the sense of enabling greater benefits, such as enhancing extreme motion of membrane components important in vision and bioenergetics. One of the scientific hurdles yet to be overcome in the membrane field involves the lack of genetically stable mutants in fatty acid pathways, mutants that seem necessary in constructing a series of chemically defined or designer membranes needed to test membrane-based theories of aging. For example, the EPA recombinant of *E. coli* used in our previous research (Valentine and Valentine, 2009) was lost after several years of culturing because unknown kinds of suppressor mutations partially restored the wild-type fatty acid composition. Indeed, the *E. coli* fabB⁻ mutant used by Metz to construct the EPA recombinant is genetically unstable due to numerous classes of suppressors occurring during routine culturing in the laboratory. Mutations partially restoring a wild-type fatty acid phenotype occur in spite of the presence of strong selective pressure applied by culturing cells in the presence of high levels of the antibiotic corresponding to the antibiotic resistance gene carried by the transposon system used to construct the mutant.

We have found that the presence of the EPA plasmid helped to stabilize the fabB⁻ mutant, allowing us to maintain the EPA recombinant for several years. This involved monthly transfers to fresh media before suppressor mutations finally occurred. In another series of experiments involving a set of cold-sensitive mutants of a marine *Shewanella* sp. selected to be dependent on polyunsaturated fatty acids for growth at 4°C (Valentine and Valentine, 2004), we found that suppressor mutations occurred frequently during growth of shake cultures at 4°C. We observed that the original EPA-plus phenotype was not restored in cultures overrun by mutants able to grow at 4°C. Instead, a new membrane composition featuring high levels of monounsaturated molecular species of phospholipids (18:1 and 16:1) permitting growth at cold temperatures was identified.

In further physiological analysis of cold-sensitive mutants of *Shewanella* some bizarre classes of fatty acid-dependent mutants were noted. In some cases specific unsaturated fatty acids restored growth at 4°C but were not required at 30°C, in contrast to other mutants in which a specific unsaturated fatty acid restored growth at 4°C but inhibited growth at 30°C. We offer no biochemical explanation for these mutants, but were struck by the idea that virtually any membrane fatty acid composition phenotype can be constructed using simple selections in the laboratory. Because these fatty acid mutants were not characterized at the genomic level, it is not possible to distinguish whether a regulatory versus a structural mutation is involved. We suggest that mutational analysis of polyunsaturated fatty acid-requiring mutants of bacteria would pay big dividends as surrogates of polyunsaturated mitochondrial membranes important in aging.

3.8 SUMMARY

Data from Bacteria and Archaea demonstrate how membrane lipids contribute to cellular bioenergetics. Membranes play fundamental roles in both production and safeguarding or conservation of cellular energy supplies. The energy produced or lost at the membrane lipid level can make the difference between life and death of bacteria. Na^+ bioenergetics has evolved as a mechanism used by many bacteria to gain energy when their membranes are chronically proton leaky. Increasing temperature or increasing fatty acid unsaturation levels tends to increase rates of energy production balanced by increased rates of futile cycling of protons. Dysfunctional conformational dynamics of fatty acid chains dictated by the environment, along with oxidatively mediated damage, are at the heart of futile proton tunneling in many prokaryotes and reach a peak in membranes enriched with DHA/EPA. As a result, DHA/EPA bacteria appear to avoid proton leakage by a strict lifestyle involving cold marine conditions along with a switch to Na^+ bioenergetics. Data from DHA/EPA bacteria have led to a unified concept of membrane biochemistry highlighting three parameters: membrane motion, cation permeability, and oxidative stability. These three essential membrane functions define the tripartite membrane fatty acid blending code, which we suggest provides a foundation for understanding the pivotal roles played by highly polyunsaturated mitochondrial membranes during aging.

REFERENCES

De Vrij, W., R. A. Bulthuis, and W. N. Konings. 1988. Comparative study of energy-transducing properties of cytoplasmic membranes from mesophilic and thermophilic *Bacillus* species. *J. Bacteriol.* 170:2359–66.

DeLong, E. F., and A. A. Yayanos. 1986. Biochemical function and ecological significance of novel bacterial lipids in deep-sea prokaryotes. *Appl. Environ. Microbiol.* 51:730–37.

Jastroch, M., A. S. Divakaruni, S. Mookerjee, et al. 2010. Mitochondrial proton and electron leaks. *Essays Biochem.* 47:53–67.

Methé, B. A., K. E. Nelson, J. W. Deming, et al. 2005. The psychrophilic lifestyle as revealed by the genome sequence of *Colwellia psychrerythraea* 34H through genomic and proteomic analyses. *Proc. Natl. Acad. Sci. USA* 102:10913–18.

Metz, J. G., P. Roessler, D. Facciotti, et al. 2001. Production of polyunsaturated fatty acids by polyketide synthases in both prokaryotes and eukaryotes. *Science* 293:290–93.

Nichols, J., and D. Deamer. 1980. Net proton-hydroxyl permeability of large unilamellar liposomes measured by an acid-base titration technique. *Proc. Natl. Acad. Sci. USA* 70:2038–42.

Tchernov, D., M. Y. Gorbunov, C. de Vargas, et al. 2004. Membrane lipids of symbiotic algae are diagnostic of sensitivity to thermal bleaching in corals. *Proc. Natl. Acad. Sci. USA* 101:13531–35.

Valentine, D. L. 2007. Adaptations to energy stress dictate the ecology and evolution of the Archaea. *Nat. Rev. Microbiol.* 5:316–23.

Valentine, R. C., and D. L. Valentine. 2004. Omega-3 fatty acids in cellular membranes: a unified concept. *Prog. Lipid Res.* 43:383–402.

Valentine, R. C., and D. L. Valentine. 2009. *Omega-3 fatty acids and the DHA principle.* Boca Raton, FL: Taylor and Francis Group.

van de Vossenberg, J. L. C. M., A. J. M. Driessen, M. S. da Costa, et al. 1999. Homeostasis of the membrane proton permeability in *Bacillus subtilis* grown at different temperatures. *Biochem. Biophys. Acta* 1419:97–104.

van de Vossenberg, J. L. C. M., A. J. M Driessen, and W. N. Konings. 1998. The essence of being extremophilic: the role of the unique archaeal membrane lipids. *Extremophiles* 2:163–70.

Yayanos, A. A., S. Ferriera, J. Johnson, S. Kravitz, A. Halpern, K. Remington, K. Beeson, B. Than, Y.-H. Rogers, R. Freidman, and J. C. Venter. *Moritella* sp. PE36. Submitted June 2007 to the EMBL/GenBank/DDBJ databases.

Yazawa, K. 1996. Production of eicosapentaenoic acid from marine bacteria. *Lipids* 31:297–300.

Section II

Darwinian Selection of Membranes Enabling Longevity

A tipping point in evolution occurred some 400 million years ago when oxygenation of the atmosphere allowed animal cells to capture ten-fold more energy than was previously possible in a world without oxygen. This great leap forward in bioenergetic capacity carried great risk because it exposed cells to the powerful chemical toxicity of derivatives of oxygen (reactive oxygen species (ROS)), which can inflict great damage and readily kill cells and organisms by an oxidative chain reaction mechanism. Data from a hierarchy of life-forms, from nematodes to whales, show that oxygen-dependent life-forms have evolved a remarkable number of strategies for combating or defending against oxidative damage. These data suggest that protecting polyunsaturated membranes against oxidative damage is a universal secret of longevity.

4 Protective Mechanisms for EPA Membranes in *C. elegans* and Their Relationship to Life Span

Caenorhabditis elegans is one of the premier model organisms for research on aging, and discoveries from this tiny nematode have helped lay a solid foundation in understanding the molecular biology of aging. *C. elegans* produces one of the most highly unsaturated membranes in nature, especially at low temperatures, where up to one-third of total membrane fatty acids are eicosapentaenoic acid (EPA). Docosahexaenoic acid (DHA) is not produced, and when incorporated in the diet, DHA is mostly retroconverted to EPA. *C. elegans* has evolved a desaturase pathway for synthesis of EPA, and mutational and regulatory analysis has established each step of its biosynthesis. A cold-activated enzyme system has evolved that generates a hierarchy of di-unsaturated molecular species of phospholipids with 10, 9, 8, 6, 5, 4, 3, and 2 double bonds. Mitochondrial membranes of *C. elegans* contain almost twice the level of double bonds as human mitochondria.

In contrast to its other membranes, mitochondrial membranes of *C. elegans* grown at 20°C contain arachidonic acid (ARA) (20:4) but little EPA. EPA-minus mutants survive but display significantly lower rates of firing of their powerful bending muscles. Rates of bending are restored to normal by feeding purified EPA or even a mixture of fish oils found in omega-3 capsules packaged for human consumption. EPA has been found to maximize rates of synaptosomal cycling at the level of endocytosis. Recently, *C. elegans* has been used as a model to test the membrane peroxidation hypothesis of aging (Hulbert et al., 2007; Shmookler Reis et al., 2011; Hulbert, 2011; also see Chapter 6). These data are highlighted below along with novel mechanisms used by *C. elegans* to protect its highly unsaturated fatty acid (HUFA)-enriched membranes against oxidative damage.

4.1 FEEDING AN EXCESS OF EPA DECREASES LIFE SPAN OF *C. ELEGANS*

In 2004 we described the properties of an EPA-minus mutant of a marine *Shewanella* strain that required EPA for growth at 4°C (Valentine and Valentine, 2004). At 30°C the EPA-requiring mutant grew without feeding EPA or other polyunsaturated chains. DHA also supported growth at 4°C, but fatty acid analysis showed that a majority of DHA was apparently enzymatically retroconverted to EPA before incorporation into membrane phospholipids. Fatty acid requirements were tested using a simple growth assay involving spreading an aliquot consisting of a few microliters of EPA or other fatty acids dissolved in ethanol along with an inoculant of about 100 μl of the *Shewanella* mutant on the surface of Petri dishes. Growth effects of unsaturated fatty acids could be scored visually or quantitatively after suspending cells with fresh medium and measuring their optical densities.

It was soon clear that growth response of cells supplemented with EPA or DHA followed a bell-shaped concentration curve, apparently reflecting both benefits and risks with respect to fatty acid concentration. Growth initially increased in a linear fashion, reaching a peak in the range of 40 to 60 μg of EPA per Petri dish surface. Beyond 60 μg of EPA per dish growth steadily declined, with about 75 percent growth inhibition reached at 100 μg of EPA or DHA per dish. These data gathered in the mid-1990s remained buried in a lab notebook until 2004, when they were mentioned in our review (Valentine and Valentine, 2004). We suggested that an excess of DHA/EPA might be toxic to some bacteria, including *E. coli*, and speculated that a toxicity test might be developed using an animal model (Valentine and Valentine, 2009). Interestingly, Harman experimented with dietary levels of fish oil in rats and concluded that HUFAs are pro-aging (see historical perspective in Chapter 1).

Recently, Shmookler Reis and colleagues (2011) reported that dietary EPA shortens the life span of *C. elegans* (Figure 4.1). As background, ecological data show that *C. elegans* forages largely for bacterial food in soil detritus. But, it is unlikely that soil bacteria taken as food by *C. elegans* produce significant amounts of EPA or DHA. The growth inhibition by EPA reported by Shmookler Reis and coworkers might be explained by the lack of the need for a suitable regulatory system for monitoring dual fluxes of EPA coming from both the diet and biosynthesis. In other words, the absence of a regulatory system might explain why excessive membrane enrichment with dietary EPA occurs. The absence of such a safety valve in *C. elegans* is not surprising given the apparent paucity of EPA in its diet in nature.

For comparison, the response of mammals to high levels of EPA/DHA in their diets can be divided into two modes—heavy enrichment of membranes in many tissues versus selective or little enrichment of membranes with these highly unsaturated chains. Among mammals, mice, which feed largely on plant-derived foods missing preformed EPA/DHA, show enrichment of their membranes when fed these fatty acids. In contrast, it is difficult to imagine a scenario in humans in which a high-EPA/DHA diet could significantly change bulk membrane fatty acid compositions. A caveat to this general rule in humans is that levels of DHA/EPA in membranes of red blood cells do reflect levels of these fatty acids in the diet. A few other cells and tissues seem to be enriched with dietary DHA/EPA. It is interesting to speculate that in future studies numerous cases of DHA/EPA toxicity across a range of organisms will be discovered. The toxic effects of DHA as an assassin of senescing colon epithelial cells, whose survival might lead to polyps and eventually cancer, were reviewed earlier (Valentine and Valentine, 2013).

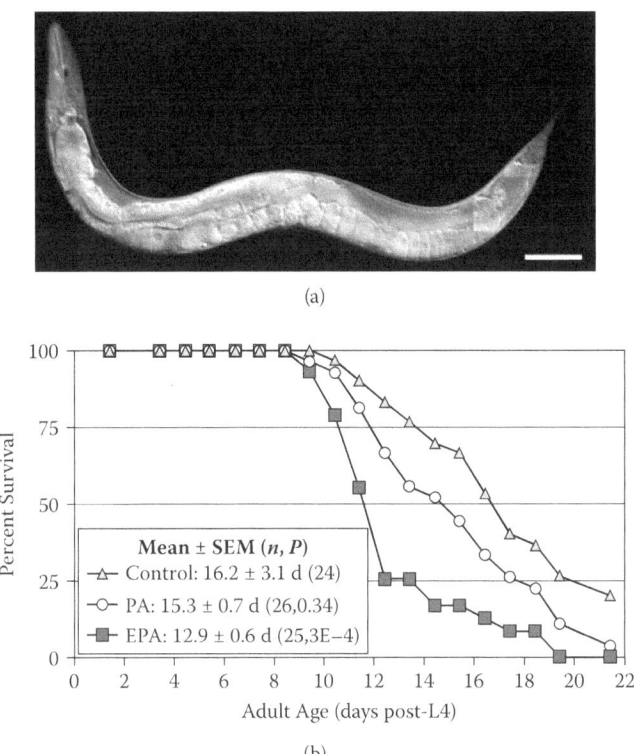

(a)

(b)

FIGURE 4.1 Life span of *C. elegans* is decreased by dietary EPA. (a) *C. elegans* shown in this photo is a voracious feeder using its powerful and rapid firing bending muscles to graze on bacteria in soil detritus. Soil bacteria lack EPA/DHA, and *C. elegans* has evolved a complete biosynthetic pathway for EPA, essential for rapid firing of bending muscles and other functions. (Photo courtesy of Dr. Ian Chin-Sang, Queen's University.) (b) Feeding EPA in combination with EPA produced by biosynthesis is toxic and shortens the life span of *C. elegans*. (Shmookler Reis et al., *Aging (Albany NY)* 3:125–47, 2011. Copyright © 2011 Shmookler Reis et al.)

4.2 *C. ELEGANS* SYNTHESIZES ONE OF THE MOST HIGHLY UNSATURATED MEMBRANES IN NATURE

C. elegans produces and enriches its membranes with high levels of the omega-3 fatty acid EPA, the sister molecule of DHA. EPA, like DHA, likely contributes extraordinary properties to membranes, including extreme motion, a tendency to leak cations, and extreme sensitivity to peroxidation (Valentine and Valentine, 2009). *C. elegans* synthesizes all of its EPA, and as mentioned above, it is unlikely that a significant level of dietary EPA is available. EPA levels increase sharply as growth temperatures drop from 25°C to 15°C, reaching levels more than 30 percent of total fatty acids in cells adapted to grow at 15°C (Tanaka et al., 1996, 1999). *C. elegans* lacks the final elongase enzyme system to generate DHA but incorporates some DHA when fed in the diet. We interpret these data to mean that compared to DHA, EPA is more beneficial to *C. elegans*, but that benefits are dependent on temperature.

Molecular biology of EPA and toxicity of dietary EPA in membranes of *C. elegans* can be explained by the nature of molecular species of EPA phospholipids present in different membranes and their effect on membrane biochemistry and lipid dynamics. In membranes of many organisms EPA is located in the sn-2 position of phospholipids, often paired with a saturated chain such as 18:0 (stearic acid) in the sn-1 position. Analysis of molecular species of phospholipids in membranes shows that *C. elegans* has evolved an enzyme system up-regulated by cold temperatures that catalyzes the formation of di-EPA in which EPA chains occupy both acyl positions on the phospholipid's glycerol headgroups (Tanaka et al., 1999). This molecular species is among the most highly unsaturated membrane phospholipids found in nature, second only to di-DHA, which is found in high levels in rhodopsin membrane disks of rod cells of the eye. Di-DHA enables rapid motion of rhodopsin, which fires the trigger reaction for vision. Di-DHA has a total of 12 (C-C) double bonds per molecule, compared to ten for di-EPA. Rhodopsin membrane disks are among the most oxidatively unstable found in nature (see Chapter 12), suggesting that EPA-enriched membranes of *C. elegans* are also major targets of oxidative damage.

Analysis of molecular species of phospholipids in *C. elegans* shows that in addition to relatively high levels of di-EPA, a virtually complete suite of di-unsaturated phospholipids is present. Presumably this versatility in molecular species of phospholipids displaying a wide range of unsaturation allows *C. elegans* to fine-tune the motional and other biochemical properties of its membrane.

4.3   EPA BENEFITS *C. ELEGANS* BY ENABLING EXTREME MEMBRANE MOTION IMPORTANT FOR RAPID CYCLING OF THE SYNAPTIC VESICLES ESSENTIAL FOR FAST FIRING OF BENDING MUSCLES

C. elegans is a voracious feeder depending on its fast-firing bending muscles for motion as it encounters a relatively large domain while grazing on bacteria. Analysis of knockout mutations targeting EPA synthesis genes has led to the finding that EPA maximizes neuromuscular firing rates of its powerful bending muscles. Wild-type *C. elegans* bends about three times per second in contrast to an EPA-minus mutant at about one bend per second (Watts and Browse, 2002). Supplementation of the growth medium with EPA restores rates of bending to levels approaching that of wild type (Watts and Browse, 2002; Lesa et al., 2003). Lack of EPA is not lethal, and mutants carrying this defective gene can be grown under laboratory conditions because they are still able to produce other di-polyunsaturated molecular species of phospholipids, such as di-18:3, acting as surrogates of EPA phospholipids. However, in its natural environment an EPA-minus mutant is unlikely to survive. We were able to confirm earlier results showing that DHA, while not normally produced by *C. elegans*, nevertheless substitutes for EPA when present in the diet. However, considerable retroconversion of DHA back to EPA occurs, confirming results by Lesa et al. (2003). One explanation of why *C. elegans* converts DHA → EPA is that EPA is significantly more beneficial than DHA in neurons and other cells of *C. elegans*. Also note that an aliquot of fish oil concentrate taken directly from omega-3-enriched fish oil capsules packaged for human consumption also restores faster rates of bending (Valentine and Valentine, 2009). DHA, EPA, and arachidonic acid (20:4) are all present in fish oil capsules. Thus, studies with *C. elegans* help establish the essential neuromuscular functions contributed by fish oils and point toward a vital and perhaps universal role or roles in neuron biochemistry.

The fast-firing bending muscles of *C. elegans* serve as a model for fast-firing muscles in other animals, including cardiac muscles of tiny mammals, breast muscles of hummingbirds, rattle muscles of rattlesnakes, and escape muscles of krill. As introduced above, rates of neuromuscular firing of bending muscles of *C. elegans* are maximized by enrichment of membranes with EPA. This EPA effect has recently been traced to the level of endocytosis of synaptic vesicles (Marza et al., 2008). These data show that rates of endocytosis of synaptic vesicles become rate limiting for muscle bending in EPA-minus mutants of *C. elegans*. Feeding EPA or ARA, the latter of which is incorporated directly into membrane phospholipids or indirectly after conversion to EPA, restores normal rates of endocytosis. At the biochemical level, presynaptic sites where synaptosomal membranes are being recycled by endocytosis have been found to have abnormally low levels of the endocytic enzyme phosphoinositide phosphatase (synaptojanin). The main synaptojanin substrate, phosphatidylinositol-4,5-bisphosphate, accumulates at these sites. Membrane lipid dynamics of endocytic membrane sites seem to explain these data. That is, these data are consistent with EPA acting to increase rates of endocytosis by enhancing collisions between synaptojanin and its phospholipid substrate. *C. elegans* grows over a relatively narrow range of temperatures from about 15 to 30°C. We suggest that EPA might be especially important for rapid firing of bending muscles at the low end of its temperature range.

A collision model for explaining the role of EPA in endocytosis of synaptic vesicles of *C. elegans* is based on the following points:

- Synaptojanin is not initially localized or enriched in membranes forming clathrin-coated pits, an essential early stage of endocytosis of synaptosomes, but is recruited and moves laterally across the membrane from nearby presynaptic membrane surfaces.
- It is proposed that DHA/EPA contributes extreme membrane motion enabling extreme membrane mobility of endocytic components. Increasing the collision rate of synaptojanin, sparsely distributed on the presynaptic membrane, with its membrane-bound inositol phosphate substrates is proposed to enable rapid rates of synaptosomal cycling.
- *C. elegans* has also evolved a unique mechanism for selective incorporation of EPA and other polyunsaturated fatty acids (PUFAs) directly into inositol phosphate substrates, presumably for the purpose of increasing the

rates of collisions between membrane-bound inositol phosphate substrates and synaptojanin.
- In essence, the mode of action of DHA/EPA in endocytosis of synaptic vesicles is envisioned to maximize collisions among critical membrane components in a manner similar to that of the trigger reactions for vision and other sensory systems (see Valentine and Valentine, 2013).

Recently Lee and coworkers (2008) have discovered a new mechanism of lipid dynamics in membranes of *C. elegans*. As mentioned above, EPA is enriched in phosphatidylinositol, a membrane-bound phospholipid that acts as a powerful hormone in *C. elegans* and many other organisms. Phosphatidylinositol enriched with acyl chains of EPA is synthesized by *C. elegans* and is expected to display extreme motion in membranes, perhaps representing another mechanism to speed up rates of endocytosis and cycling of synaptic vesicles in *C. elegans* and other animals.

Marza and colleagues (2008) reviewed the contribution of the membrane to endocytosis of synaptic vesicles and conclude that highly unsaturated fatty acid chains are likely universally important in organisms ranging from *C. elegans* to humans. This conclusion is based in part on data from Zimmer and colleagues (2000), who showed that rats chronically deprived of highly unsaturated fatty acids in the diet display a decrease in numbers of synaptosomes of dopaminergic neurons.

4.4   SENSORY PERCEPTION IN *C. ELEGANS* LIKELY DEPENDS ON EPA ENABLING EXTRAORDINARY MEMBRANE MOTION

Remarkable progress has been made in understanding the molecular biology of sensory perception in *C. elegans* (see review by Bargmann, 2006), opening a new window for understanding essential roles of omega-3s in sensory membranes. This tiny worm has evolved specialized sensory neurons often tipped with cilia whose membranes are packed with receptors for sensing many odorants (both water soluble and volatile), mechanical cues, oxygen levels, toxic chemicals such as copper, and even light. Bargmann (2006) concludes that these specialized membranes of neurons allow *C. elegans* to sense its external and internal environment using at least five different sensory perception systems as follows:

- Chemosensation
- Mechanosensation
- Osmosensation
- Thermosensation
- Photosensation

Bargmann (2006), in her comprehensive review of chemosensation in *C. elegans*, points out that as many as 500 to 1000 different G protein-coupled receptors are expressed in membranes of chemosensory neurons of *C. elegans*. The rhodopsin (photoreceptor)-transducin (G protein) system triggering the visual cascade in the human eye is a classic example of a sensory receptor–G protein system. *C. elegans* does not have eyes but is able to sense sunlight via a gustatory receptor modified to directly or indirectly detect the near-ultraviolet spectrum of sunlight (Edwards et al., 2008). Liu and colleagues (2010) have found that photosensation in *C. elegans* involves a taste receptor homolog (LITE-1) coupled with a G protein-dependent pathway, the latter similar in general design to the trigger reaction for vision in the eye. There is another similarity between the mechanisms of photosensation in *C. elegans* and phototransduction of light signals in photoreceptor cells of the eye. Specifically, the taste receptor homolog of *C. elegans*, which acts as a light sensor, displays eight transmembrane segments, and thus likely depends on collisions with membrane-embedded G proteins to trigger the light sensation cascade. We suggest that light sensation as well as other sensory perception systems of *C. elegans* depend on membrane lipid dynamics enabled by conformational dynamics of EPA and other polyunsaturated chains to modulate rates of collisions between sensory receptors and their G proteins (see Valentine and Valentine, 2009, 2013, for further discussion on how the wild conformational shifts of DHA are harnessed in vision and other membrane processes). In addition to sensory perception and endocytosis, we propose that HUFA-mediated membrane dynamics help maximize energy production and energy efficiency in energy-transducing membranes of mitochondria of *C. elegans*.

4.5   SENSORY SYSTEMS THAT PROTECT AGAINST MEMBRANE OXIDATION AND OXIDATIVE STRESS

The following sensory systems linked to negative taxis likely evolved to help protect membranes of *C. elegans* against autoxidation and photooxidation:

- O_2
- Copper
- Heat
- Sunlight

As background, the chemical stability of biomembranes determining the per-oxidation rates in an oxygenated environment is dependent to a large degree on the number of double bonds in the fatty acid chains of phospholipids. The chemical principles governing lipid peroxidation are now understood in detail and are readily applied to mitochondrial and neural membranes of *C. elegans*. The oxidation of unsaturated fatty acids is a spontaneous chemical process and, as such, will occur wherever the proper conditions exist, from a bottle of cooking oil to a membrane of *C. elegans*. Polyunsaturated fatty acids are far more susceptible to oxidation than monounsaturated fatty acids, with linolenic acid (18:2n6) being oxidized at a rate more than an order of magnitude greater than oleic acid (18:1n9). The large increase in oxidation rate is due to the methylene-interrupted structure of the double bonds and their facile allylic hydrogens (see Chapter 20). For every double bond past two per fatty acid, the rate of oxidation approximately doubles; thus, the rate of oxidation of DHA (22:6) is estimated to be about 480 times the rate of oxidation of 18:1n9, and the rate for EPA (20:5) is estimated to be about one-half of the DHA value. Various environmental factors, such as increased temperature, trace metals (e.g., copper and iron), oxygen concentration, and especially irradiance, can greatly increase oxidation rates. Rates of photooxidation catalyzed by visible light are much higher than with autoxidation (dark). For example, the rate of photooxidation of 18:1n9, such as that which occurs in membranes of plants, is about 30,000 times that of autoxidation, and photooxidation of 18:3n3 occurs about 900 times faster than does autoxidation (Gunstone, 1996).

C. elegans quickly moves away from bright sunlight using a unique photo-sensing cascade to trigger avoidance by negative taxis (Ward et al., 2008). The trigger mechanism introduced above involves a taste receptor homolog that apparently has evolved as a photoreceptor (LITE-1). LITE-1 is membrane bound, meaning that this protein is fixed and carries out its mode of action dependent on membrane motional dynamics, especially lateral motion. The series of steps comprising the photoreceptor cascade of *C. elegans* are light (near ultraviolet) $\rightarrow$ LITE-1 $\rightarrow$ G protein $\rightarrow$ cyclic GMP system (membrane bound) $\rightarrow\rightarrow\rightarrow$ negative taxis. This cascade occurs in membranes of light-sensitive neurons, membranes likely enriched with HUFAs, including EPA and ARA. We suggest that the sensitivity and rates of firing of this light-dependent cascade might be enabled by lipid dynamics, especially lateral motion of the membrane itself. Thus, an EPA-enriched membrane of a light-sensing neuron can, from a biochemical perspective, be considered not only to act as a physical matrix for housing sensory proteins, but also to play an indirect role in determining rates of catalysis measured in terms of frequencies of collisions among membrane components. In the language of classical chemical catalysis, this is called a solvent effect, in this case mediated by the lipid environment of the membrane itself. The discovery of photosensation in *C. elegans* raises the interesting question of how light is perceived by LITE-1 acting as a trigger protein (Diaz and Sprecher, 2011).

Several other novel mechanisms that protect EPA-enriched membranes of *C. elegans* against oxidative damage are discussed next. In addition to a transparent body that exposes its mitochondrial and HUFA-enriched membranes, including membranes of 302 neurons, to visible and ultraviolet radiation, *C. elegans* depends on O_2 for growth (Voorhies and Ward, 2000). Thus, membranes of *C. elegans* are prime targets of both photooxidative and autoxidation (peroxidation) damage due to their high unsaturation levels. As mentioned above, at 15°C *C. elegans* produces up to 34 percent of its total membrane fatty acids as EPA, with a significant fraction as di-EPA phospholipids. These structures stand out in being among the most unsaturated membrane components documented in nature (i.e., ten double bonds) and are expected to be highly vulnerable to oxidative damage. *C. elegans* has evolved an umbrella of conventional biochemical mechanisms against oxidative damage and, in addition, has evolved novel behavioral patterns that likely reduce membrane peroxidation (Gray et al., 2004; Chang et al., 2006). *C. elegans* moves away from ambient oxygen, which is apparently toxic, and seeks environments under soil detritus with subambient O_2 levels of roughly 4 to 12 percent oxygen concentrations, avoiding higher or lower concentrations. Also note that direct sunlight seldom reaches this niche, which means that worms living in a lowered O_2 environment are simultaneously protected from both autoxidation and photooxidative damage. If there is no area of low oxygen concentration to be found, the worms will congregate and lower the oxygen tension through their communal respiration. Thus, *C. elegans* uses both conventional biochemical defenses and migratory and communal mechanisms to avoid both atmospheric levels of O_2 and direct sunlight. We hypothesize that these behaviors evolved to help minimize peroxidation of EPA and other PUFAs and HUFAs whose peroxidation can trigger cellular death; we further suggest that incorporation of EPA into membranes of *C. elegans* must provide a strong selective advantage in order to force these fundamental behaviors.

Pioneering research on *C. elegans* has helped establish another important concept of aging—temperature is often a key modulator of aging, with cool temperatures favoring longevity (Van Voorhies and Ward, 1999). These authors found that longevity of wild-type *C. elegans* at 10°C was roughly four times greater than at 25°C. What is striking about these data is that long-lived nematodes generated genetically for enhanced longevity seldom surpass the increased life span seen in wild-type *C. elegans* grown at low temperatures. A clue regarding temperature modulation of aging comes from studies of mitochondria of *C. elegans*. There is an increase in organelle number and in the frequency of organelle division with an up-shift of temperature from 15 to 25°C (Labrousse et al., 1999). Rates of respiration (Van Voorhies and Ward, 1999) and ubiquinone levels (Jonassen et al., 2002) are doubled with a similar shift of temperature. Increased numbers of mitochondria and faster rates of respiration triggered by growth at 25°C might be accounted for by several possible mechanisms, including the need for more energy for growth at high temperature, increased proton uncoupling caused by high temperature, or both.

4.6 AN EXTREMELY LONG-LIVED MUTANT DECREASES EPA WHILE INCREASING LEVELS OF MONOUNSATURATED CHAINS

The dauer stage of *C. elegans* is a highly stress-resistant alternative larval stage. The dauer state is induced by starvation, a dauer pheromone, or high temperature. Dauer larvae are resistant to a variety of stresses, including the potent oxidative poison paraquat (see Chapter 2). The FOXO system (DAF-16) and the insulin-like regulatory system (ILS) of *C. elegans* are involved in induction of the dauer stage. Larvae in the dauer state are extremely long-lived compared to worms senescing in the reproductive growth phase. An extremely long-lived mutant of *C. elegans* with a life span almost ten times greater than wild type has recently been constructed. This remarkable long-lived mutant features a truncation of the *age*-1 gene upstream of the kinase domain (Ayyadevara et al., 2008). The *age*-1 gene encodes a critical protein kinase (AGE-1), which catalyzes the formation of the second messenger, phosphatidylinositol-3,4,5-trisphosphate (PIP_3). AGE-1 is part of the (ILS) regulatory cascade of *C. elegans*, well known to regulate resistance to numerous stresses, including oxidative stress. This cascade also features FOXO (DAF-16), in which partial loss of function mutations increase longevity. Allelic variation in the protein kinase P13K also appears to affect human longevity (Barbieri et al., 2003). Recent data from the laboratory of Shmookler Reis et al. (2011) show that the longevity caused by truncated *age*-1 is reflected by a reduction of EPA in membranes of the long-lived mutant. EPA enrichment of membranes dropped from about 15 percent total fatty acids typical of wild-type worms (20°C) to 6.4 percent EPA in the long-lived mutant. Levels of 20:4 also decreased in the mutant. These researchers also report that while EPA/ARA levels are decreasing, levels of monounsaturates are increasing, from around 30 percent total fatty acids in wild type to 58.6 percent in the long-lived mutant. The data of Shmookler Reis and colleagues (2011) provide important insights with likely universal application regarding regulation of longevity by membrane unsaturation. Using mutational, biochemical, regulatory, and dietary analysis, these researchers show that fatty acid unsaturation levels in C. *elegans* govern rates of membrane peroxidation and are one of the causal mechanisms contributing to life span (Figure 4.2). For instance, lowering EPA levels decreases rates of peroxidation, increasing life span. In the longest-lived mutant of C. *elegans*, EPA levels drop by more than 60 percent, while at the same time monounsaturates almost double. Note that a fatty acid chain with a single double bond is estimated to be more than two orders of magnitude more stable to peroxidation than EPA.

Increasing monounsaturates while decreasing EPA levels is proposed to contribute to longevity by decreasing membrane oxidative damage. These molecular data (Figure 4.2) are among the strongest yet published in support of and extending

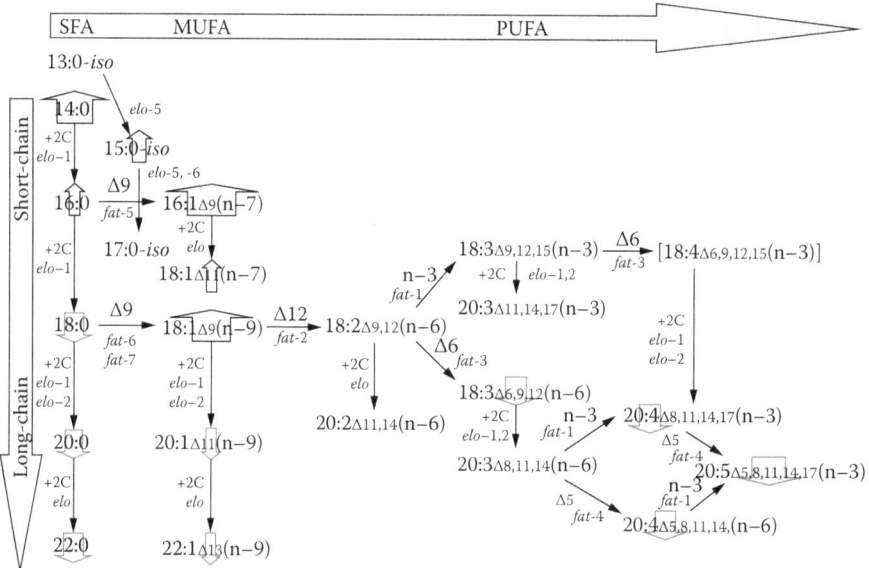

FIGURE 4.2   See color insert. Regulatory patterns of biosynthesis of EPA and other unsaturated fatty acids linked to aging. See Shmookler Reis and colleagues (2011) for a detailed explanation of regulatory patterns of fatty acid synthesis in C. *elegans*. Three points are highlighted here: (1) Black arrows show the overall pathway of membrane fatty acid biosynthesis leading to EPA. (2) Enzyme activities (+2C for elongases; Δn, n-3, or n-6 for desaturases) and their genes are shown beside each arrow. (3) Bold (dark) font over the fatty acid classes indicates up- or down-regulation with increasing life span. The width of the arrow corresponds to the strength of the correlation to longevity. (Shmookler Reis et al., *Aging (Albany NY)* 3:125–47, 2011. Copyright © 2011 Shmookler Reis et al.)

the membrane pacemaker/peroxidation theory of aging proposed by Hulbert and Pamplona, respectively (see Chapter 6 and Hulbert's discussion (2011)). We suggest that decreasing membrane peroxidation significantly lowers energy expenditure, in essence lowering energy stress.

4.7   *C. ELEGANS* TARGETS EPA AWAY FROM MITOCHONDRIAL MEMBRANES LIKELY AS A PROTECTIVE MECHANISM AGAINST OXIDATIVE DAMAGE

Cardiolipin (CL) targeted mainly to the inner mitochondrial membrane is the signature building block of mitochondria of *C. elegans* and other organisms (see Chapters 8 and 20). Recall that the inner mitochondrial membrane not only houses respiratory machinery, but also helps govern the proton permeability (i.e., energy-conserving) properties of mitochondria (see Chapter 10). CL comprises an estimated 15 percent of this total inner membrane surface and features an unusual structure in which two phospholipids are joined head to head. CL is a membrane building block with four fatty acid tails in contrast to two acyl chains of conventional membrane phospholipids. The major molecular species of CL in mitochondrial membranes of *C. elegans* grown at 20°C are structures whose acyl tails are 18:3 and 20:4 rather than 20:5, with total numbers of unsaturated double bonds ranging from 13 to 15 per CL molecule. These molecular species of CL, especially molecules enriched with 20:4 (CL-20:4), remain highly sensitive to oxidative damage due to their extraordinary numbers of double bonds. However, replacing EPA with ARA or 18:3 is estimated to halve or quarter rates of peroxidation of mitochondrial membranes, respectively.

Total CL levels in mitochondria of *C. elegans* were found to drop approximately 50 percent in older (day 10) animals compared to young (day 3) animals (Gruber et al., 2011). This is the period when ATP levels, respiration rates, and total mtDNA numbers are also dropping. These data document a significant age-dependent decline in energy metabolism and focus attention on highly unsaturated molecular species of CL as modulators of energy production and aging. CL molecular species composed of fatty acids with fewer double bonds (e.g., 18:1) are proposed to favor longevity in contrast to EPA species proposed to shorten life span (Shmookler Reis et al., 2011). CL-18:3 and CL-20:4, both being more saturated than EPA (20:5), are expected to significantly decrease oxidative stress compared to EPA-CL, an adaptation favoring longevity. However, the unsaturation level of CL of *C. elegans* even in the absence of EPA is still relatively high compared to that of longer-lived animals, including humans, and accounts for the apparent rapid, oxidatively mediated depletion of CL from mitochondrial membranes of *C. elegans* during the short life span of this worm. The main point is that EPA, which is considered to be a major target of peroxidation in membranes of *C. elegans*, is targeted away from mitochondrial membranes and instead is apparently incorporated into other classes of membranes where benefits outweigh risks. We suggest that the opposite scenario is true for mitochondrial CL where risks of EPA are proposed to be greater than benefits. Selective targeting of DHA away from membranes of mitochondria as a mechanism enabling longevity in humans is discussed in Section IV.

4.8   GROWTH OF *C. ELEGANS* REQUIRES METHYL-BRANCHED FATTY ACIDS, WHICH ARE PEROXIDATION RESISTANT

Methyl-branched fatty acids are widely distributed in soil bacteria often consumed as a food source by *C. elegans*. In bacterial membranes the extra methyl group near the methyl end of the chain is proposed to tighten the permeability barrier against protons while enhancing membrane fluidity. As an added bonus, methyl-branched fatty acids are relatively resistant to peroxidation and are considered to contribute both oxidative and energy stress tolerance to bacteria. Methyl-branched fatty acids are synthesized by birds and mammals apparently as waterproofing for hair and feathers, but are rarely found in abundance in membrane phospholipids of animals. *C. elegans* is an exception, and the complete pathway of biosynthesis has been identified; knockout mutants block synthesis, and the result is a dependence on dietary methyl-branched fatty acids. *C. elegans* requires about 5 to 10 percent of its total membrane fatty acids as methyl-branched derivatives (Kniazeva et al., 2004; Seamen et al., 2009). The precise biochemical role of methyl-branched fatty acids in membranes of *C. elegans* is not fully understood, nor is any linkage to longevity. However, replacing methyl-branched chains with straight, saturated chains has pathological effects (Seamen et al., 2009). It would not be surprising if methyl-branched chains were found to contribute to energy-saving properties. enabling longevity in *C. elegans*.

4.9   MUTATIONAL ANALYSIS OF NUCLEAR HORMONE RECEPTORS PROVIDES AN ALTERNATIVE VIEW OF THE ROLES OF LIPIDS IN REGULATING LIFE SPAN

Nuclear hormone receptors (NHRs) are ligand-modulated transcription factors that play a central role in the cell's ability to sense and respond to lipophilic signals by regulating the appropriate target genes (Pathare et al., 2012). Target genes include beta-oxidation, fatty acid desaturases, and sphingolipid biosynthesis. NHRs appear to be universal among mammals, and Pathare and colleagues (2012) dissected the roles of NHRs using *C. elegans* as a model organism. These data have led to a new perspective of the roles of membrane lipids in governing life span in *C. elegans* (Figure 4.3). For example, instead of enhancing life span, an excess of saturated fatty acids seems to decrease life span (Figure 4.3a), with a model for regulatory events shown in Figure 4.3c. One of the most provocative aspects of these data concerns the dysfunctionality of mitochondria mediated by NHR mutations. Mechanisms behind defects in mitochondria are not understood, but these authors (Pathare et al., 2012) show that both aberrant mitochondrial morphologies and decreased levels of respiration are involved. These data are consistent with the idea that NHR mutations → energy stress → aging.

We offer two suggestions regarding the interpretation of these data, both dealing with dysfunctional membrane lipid homeostasis and its effects on bioenergetics.

The first point is that an excess of saturated chains can have multiple negative effects on mitochondrial membranes and membranes in general, including increased

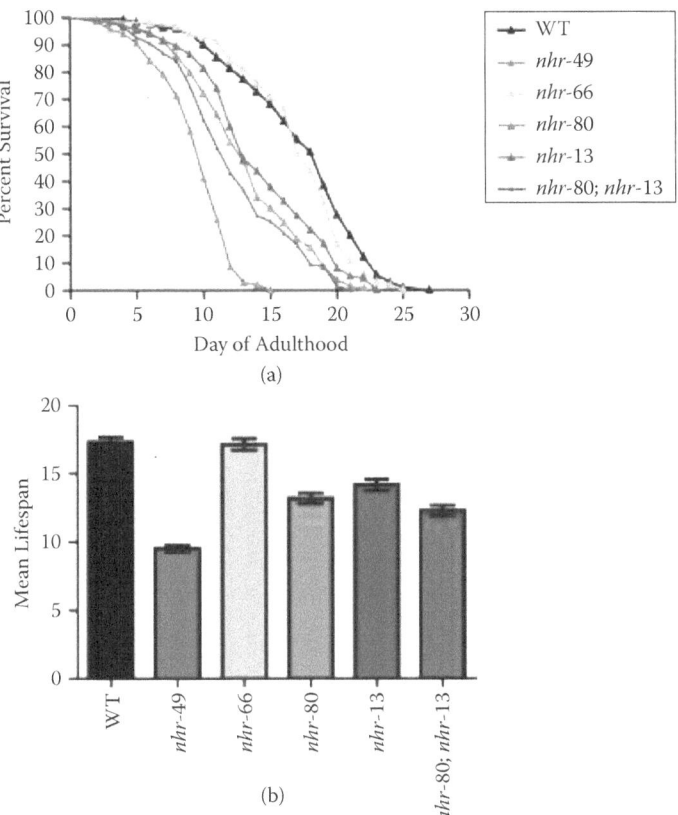

FIGURE 4.3 Nuclear hormone receptors (NHRs) modulate fatty acid synthesis, and mutations in these genes reduce life span in *C. elegans*. (a) Time course for survival. (b) The bar graph represents the combined mean adult life span (days). Note that mutations either in *nhr*-49 or its partner proteins (*nhr*-13, *nhr*-80) significantly reduce life span. *(continued)*

proton leakiness, dysfunctional mitochondrial fission, and decreased rates of electron transport, counterbalancing any benefits related to membrane oxidative stability. The second point involves combining data from Pathare and colleagues (2012) with that of Seamen and coworkers (2009). The latter researchers show that sphingolipids play essential but biochemically poorly defined roles in growth of *C. elegans*. Specifically, incorporating an excess of the saturated fatty acid, stearic acid (18:0) in place of methyl-branched fatty acids in glycosphingolipids has been proposed by Seamen and colleagues (2009) to destroy sphingolipid function and cause lethality. The data of Pathare and coworkers (2012) provide a regulatory perspective to help explain the essential roles played by membrane unsaturation, methyl-branched fatty acids, and sphingolipids in membranes of *C. elegans*.

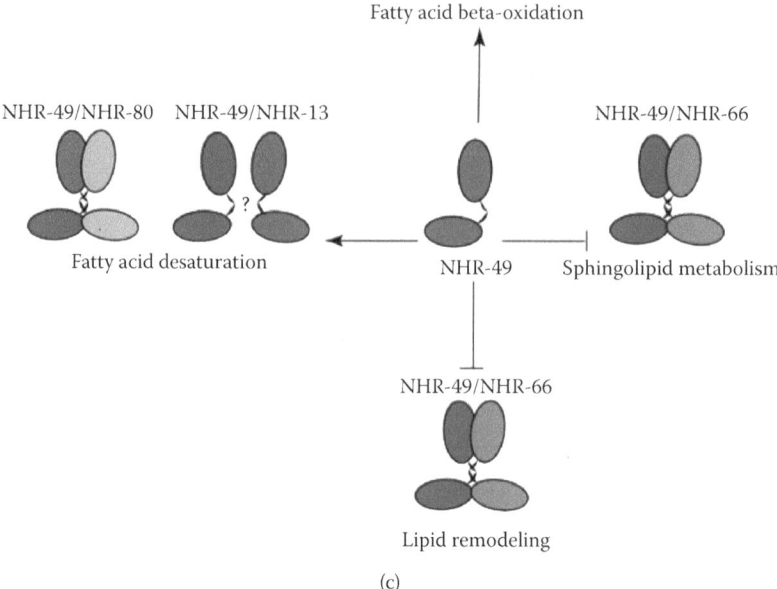

Fatty acid beta-oxidation

NHR-49/NHR-80   NHR-49/NHR-13   NHR-49/NHR-66

Fatty acid desaturation   NHR-49   Sphingolipid metabolism

NHR-49/NHR-66

Lipid remodeling

(c)

FIGURE 4.3 *(continued)* Nuclear hormone receptors (NHRs) modulate fatty acid synthesis, and mutations in these genes reduce life span in *C. elegans*. (c) Model of NHR-49-dependent regulation of lipid metabolism. According to this model, NHR-49 interacting with partner regulatory protein NHR-66 represses genes involved in sphingolipid processing and lipid remodeling. In contrast, NHR-49 binds to NHR-80 to activate the fatty acid desaturase genes. Thus, nuclear hormone receptor genes appear to regulate EPA synthesis with impact on life span. (From Pathare et al., *PLoS Genet.* 8:e1002645, 2012. doi: 10.1371/journal.pgen.1002645.)

4.10 SUMMARY

The contribution of membrane unsaturation to aging in *C. elegans* is being actively studied and likely has universal implications. We suggest that the DHA principle and tripartite membrane fatty acid blending code developed with bacteria are applicable to explain some of the surprising biochemical features of membranes of *C. elegans*. *C. elegans* has evolved a cold-activated enzyme that catalyzes the attachment of a second EPA molecule to phospholipids forming di-EPA. The phase transition temperature of di-EPA is lower than minus 50°C, making this molecular species a powerful membrane antifreeze comparable to di-DHA of rhodopsin membrane disks of the eye. Recall that rhodopsin protein has been calibrated to rotate and move laterally at blazing speed in rhodopsin membrane disks. We propose that di-EPA and similar molecular species in *C. elegans* work by maximizing motion of membrane components, as discussed above in the case of synaptic vesicle cycling. Indeed, membrane motion may be behind the evolution of another enzyme that attaches EPA to phosphatidylinositol, presumably enabling extreme motion of this membrane-embedded regulatory species of phospholipid.

The highly unsaturated membranes of *C. elegans* are unusually sensitive to chemical peroxidation, and at least five novel mechanisms have evolved to minimize damage. These include retroconversion of DHA → EPA, taxis to regions of lower oxygen, communal respiration to lower O_2 levels, selective targeting of EPA away from cardiolipin of mitochondria in favor of chains less susceptible to peroxidation, and up-regulation of ratios of 18:1/EPA. In addition, conventional mechanisms against oxidative stress, including antioxidants and enzymatic detoxification of ROS, are important.

EPA membranes of *C. elegans*, like DHA-enriched rhodopsin membrane disks of the human eye, are targets or substrates for chemical oxidation by molecular oxygen. Thus, DHA, EPA, and even polyunsaturated fatty acids in mitochondria are not always beneficial, and indeed can be considered a double-edged sword in terms of benefits versus risks. However, we propose that these substantial risks must be counterbalanced by extraordinary biochemical benefits to account for the prevalence of highly unsaturated fatty acids in cellular membranes of *C. elegans*. We further hypothesize that chemical oxidation and toxicity of peroxidation products strongly modulates the distribution pattern of DHA/EPA/polyunsaturated chains in cellular membranes of *C. elegans*. We suggest that the distribution pattern of unsaturated chains in mitochondrial membranes of *C. elegans* and other animals is honed by Darwinian selection, leading to the fittest oxidatively stable and energy-conserving membranes, enabling longevity. Finally, *C. elegans* continues to be a powerful research tool for studies of aging; for example, recent data are consistent with cardiolipin acting both as a reporter of oxidative stress and as a trigger for aging. Cardiolipin is proposed to play similar roles in humans.

REFERENCES

Ayyadevara, S., R. Alla, J. J. Thaden, et al. 2008. Remarkable longevity and stress resistance of nematode PI3K-null mutants. *Aging Cell.* 7:13–22.

Barbieri, M., M. Bonafè, C. Franceschi, et al. 2003. Insulin/IGF-I-signaling pathway: an evolutionarily conserved mechanism of longevity from yeast to humans. *Am. J. Physiol. Endocrinol. Metab.* 285:E1064–71.

Bargmann, C. I. 2006. Chemosensation in *C. elegans*. *WormBook* 25:1–29.

Chang, A. J., N. Chronis, D. S. Karow, et al. 2006. A distributed chemosensory circuit for oxygen preference in *C. elegans*. *PLoS Biol.* 4:e274.

Diaz, N. N., and S. G. Sprecher. 2011. Photoreceptors: unconventional ways of seeing. *Curr. Biol.* 21:R25–27.

Edwards, S. L., N. K. Charlie, M. C. Milfort, et al. 2008. A novel molecular solution for ultraviolet light detection in *Caenorhabditis elegans*. *PLoS Biol.* 6:e198.

Gray, J. M., D. S. Karow, H. Lu, et al. 2004. Oxygen sensation and social feeding mediated by a *C. elegans* guanylate cyclase homologue. *Nature* 430:317–22.

Gruber, J., L. F. Ng, S. Fong, et al. 2011. Mitochondrial changes in ageing *Caenorhabditis elegans*—what do we learn from superoxide dismutase knockouts? *PLoS One.* 6:e19444.

Gunstone, F. D. 1996. *Fatty acid and lipid chemistry*. London: Blackie Academic & Professional, p. 252.

Hulbert, A. J. 2011. Longevity, lipids and *C. elegans*. *Aging (Albany NY)* 3:81–82.

Hulbert, A. J., R. Pamplona, R. Buffenstein, and W. A. Buttemer. 2007. Life and death: metabolic rate, membrane composition, and life span of animals. *Physiol. Rev.* 87:1175–213.

Jonassen, T., B. N. Marbois, K. F. Faull, et al. 2002. Development and fertility in *Caenorhabditis elegans* clk-1 mutants depend upon transport of dietary coenzyme Q8 to mitochondria. *J. Biol. Chem.* 277:45020–27.

Kniazeva, M., Q. T. Crawford, M. Seiber, C.-Y. Wang, and M. Han. 2004. Monomethyl branched-chain fatty acids play an essential role in *Caenorhabditis elegans* development. *PLoS Biol.* 2(9):e257.

Labrousse, A. M., M. D. Zappaterra, D. A. Rube, et al. 1999. *C. elegans* dynamin-related protein DRP-1 controls severing of the mitochondrial outer membrane. *Mol. Cell.* 4:815–26.

Lee, H. C., T. Inoue, R. Imae, et al. 2008. *Caenorhabditis elegans* mboa-7, a member of the MBOAT family, is required for selective incorporation of polyunsaturated fatty acids into phosphatidylinositol. *Mol. Biol. Cell.*19:1174–84.

Lesa, G. M., M. Palfreyman, D. H. Hall, et al. 2003. Long chain polyunsaturated fatty acids are required for efficient neurotransmission in *C. elegans. J. Cell Sci.* 116:4965–75.

Liu, J., A. Ward, J. Gao, et al. 2010. *C. elegans* phototransduction requires a G protein-dependent cGMP pathway and a taste receptor homolog. *Nat. Neurosci.* 13:715–22.

Marza, E., T. Long, A. Saiardi, M. Sumakovic, S. Eimer, et al. 2008. Polyunsaturated fatty acids influence synaptojanin localization to regulate synaptic vesicle recycling. *Mol. Biol. Cell.* 19:833–42.

Pathare, P. P., A. Lin, K. E. Bornfeldt, et al. 2012. Coordinate regulation of lipid metabolism by novel nuclear receptor partnerships. *PLoS Genet.* 8:e1002645. doi: 10.1371/journal.pgen.1002645.

Seamen, E., J. M. Blanchette, and M. Han. 2009. P-type ATPase TAT-2 negatively regulates monomethyl branched-chain fatty acid mediated function in post-embryonic growth and development in *C. elegans. PLoS Genet.* 5:e1000589.

Shmookler Reis, R. J., L. Xu, H. Lee, et al. 2011. Modulation of lipid biosynthesis contributes to stress resistance and longevity of *C. elegans* mutants. *Aging (Albany NY)* 3:125–47.

Tanaka, T., K. Ikita, T. Ashida, et al. 1996. Effects of growth temperature on the fatty acid composition of the free-living nematode *Caenorhabditis elegans. Lipids* 31:1173–78.

Tanaka, T., S. Izuwa, K. Tanaka, et al. 1999. Biosynthesis of 1,2-dieicosapentaenosyl-sn-glycero-3-phosphocholine in *Caenorhabditis elegans. Eur. J. Biochem.* 263:189–94.

Valentine, R. C., and D. L. Valentine. 2004. Omega-3 fatty acids in cellular membranes: a unified concept. *Prog. Lipid Res.* 43:383–402.

Valentine, R. C., and D. L. Valentine. 2009. *Omega-3 fatty acids and the DHA principle.* Boca Raton, FL: Taylor and Francis Group.

Valentine, R. C., and D. L. Valentine. 2013. *Neurons and the DHA principle.* Boca Raton, FL: Taylor and Francis Group.

Van Voorhies, W. A., and S. Ward. 1999. Genetic and environmental conditions that increase longevity in *Caenorhabditis elegans* decrease metabolic rate. *Proc. Natl. Acad. Sci. USA* 96:11399–403.

Van Voorhies, W. A., and S. Ward. 2000. Broad oxygen tolerance in the nematode *Caenorhabditis elegans. J. Exp. Biol.* 203:2467–78.

Ward, A., J. Liu, Z. Feng, et al. 2008. Light-sensitive neurons and channels mediate phototaxis in *C. elegans. Nat. Neurosci.* 11:916–22.

Watts, J. L., and J. Browse. 2002. Genetic dissection of polyunsaturated fatty acid synthesis in *Caenorhabditis elegans. Proc. Natl. Acad. Sci. USA* 99:5854–59.

Zimmer, L., S. Delpal, D. Guilloteau, et al. 2000. Chronic n-3 polyunsaturated fatty acid deficiency alters dopamine vesicle density in the rat frontal cortex. *Neurosci. Lett.* 284:25–28.

5 Remarkable Longevity of Queens of Social Insects Likely Involves Dietary Manipulation to Minimize Levels of Polyunsaturates and Decrease Membrane Peroxidation

Genomics of social insects has come of age (Honeybee Genome Sequencing Consortium, 2006; Smith et al., 2011a, 2011b; Wurm et al., 2011). These data show the extraordinary adaptability of bees and ants with respect to social habits, invasiveness, biochemistry, olfaction, communication, reproduction, resistance to biological, chemical, and physical stresses, and longevity. These data also show why social insects deserve their elite title of super organisms.

Interest here, though narrowly focused on longevity of insects (Pamilo, 1991; Schrempf et al., 2011; Schneider et al., 2011; Keller and Jemielity, 2006; Keller and Genoud, 1997), benefits from the diversity of topics covered in the outstanding genomics papers above. These data provide a unique perspective of how longevity is a dynamic and evolvable trait and provide a picture of how readily longevity can be manipulated to fit the lifestyle of insects. Genomic analysis of patterns of lipid metabolism in social insects shows the presence of an extraordinary large number of lipid genes encoding multiple fatty acid desaturation pathways. Note that terrestrial insects seldom exceed four double bonds per membrane fatty acid chain, in contrast to aquatic insects, which often enrich their membranes with highly unsaturated fatty acids (HUFAs), especially eicosapentaenoic acid (EPA). Genes enabling rapid buildup and breakdown of storage lipids have been identified. Genomic analysis provides a framework for understanding the extreme longevity of queens of social insects, with emphasis here on how dietary manipulation of fatty acids enables longevity.

There are a number of clues in the literature that point toward an important relationship between membrane unsaturation and the extraordinary long life span of queen bees and ants. It has been known for many years that larvae of bees fed only royal jelly differentiate into queens, in contrast to larvae destined to be workers, which are fed for only two days with royal jelly, then switched to a diet including pollen. Thus, diet is believed to be largely responsible for the extreme longevity of queens, and conversely the ten- to twenty-fold shorter life span of workers. A 57 kDa protein in royal jelly, royalactin, has recently been found to induce a sophisticated cascade of differentiation and biochemistry characteristic of queens of social insects. Royalactin behaves as a key signaling molecule, causing gigantism and robust ovary development essential for queens. Thus, the cascade for extreme longevity in social insects now has a dietary trigger molecule, royalactin, and an end point, remarkable longevity. Obviously, there are many steps between the trigger and longevity, and the purpose of this chapter is to illuminate the membrane's contribution to aging and longevity in insects. Social insects, including honeybees and ants, are highlighted along with *Drosophila*, the latter a well-established model organism for research on aging.

5.1   ROYALACTIN

Queens of social insects, including bees and ants, stand out from workers in their giant size, prolific and long-term egg-laying capacity, and extreme longevity. Queen bees may live ten times longer than workers, with queens of certain species of ants living for greater than twenty years (i.e., about twenty times longer than workers). It has long been suspected that royal jelly fed to queens for their entire life span somehow is responsible for differentiation and longevity. Now in an important milestone in the field of dietary control of longevity, it has recently been reported that royalactin, a 57 kDa protein in royal jelly, induces differentiation of honeybee larvae into queens (Kamakura, 2011).

Royalactin increases body size and ovary development and shortens developmental time in honeybees. Surprisingly, it also shows similar effects in the fruit fly, *Drosophila*. Royalactin activates a key regulatory enzyme, p7056 kinase, which is responsible for the increase of body size. Royalactin also increases the titer of juvenile hormone, an essential hormone for ovary development. Knockdown of epidermal growth factor receptor (Egfr) expression in the fat body of honeybees and fruit flies results in a defect of all phenotypes induced by royalactin. These data show that Egfr mediates these actions. These findings indicate that royalactin drives queen development through an Egfr-mediated signaling pathway and opens the door toward understanding dietary-mediated longevity of queens. A closer look at the composition of royal jelly suggests that other components (or lack of components) also contribute to longevity of queens of honeybees (Haddad et al., 2007) and likely queens of other social insects as well.

5.2   ROYAL JELLY

Royal jelly contains the basic metabolic building blocks needed by the queen for her production of a steady stream of eggs needed to maintain a healthy hive (Lercker et al., 1981; Melliou and Chinou, 2005). Royal jelly is a creamy, white secretion made in hypopharyngeal and mandibular glands of young worker bees aged between five and fifteen days. This secretion is fed for a maximum of three days to the brood of workers and drones in the beehive. In contrast, royal jelly constitutes the exclusive food of the queen for her entire life span, both her larval and adult lives. Pollen collected by workers, and likely some honey, is altered dramatically in composition during metabolism and conversion to royal jelly. For example, species of polyunsaturated fatty acids, molecules that are a signature of plant pollen, are virtually eliminated with a few long-chain length fatty acids suitable for incorporation into the membranes remaining. Indeed, the major lipid component of royal jelly is a mixture of novel fatty acids of eight to ten carbon atoms, chains that are unsuitable or even inhibitory as building blocks for membranes. Thus, the queen throughout her life likely synthesizes all of her membrane fatty acids de novo with relatively little input from dietary fatty acids. The end result is that cellular membranes of the queen are markedly different in fatty acid composition, especially when compared to mature workers that fly for a living. The chemical composition of royal jelly of bees is well known and consists of various classes of nutrients that support the growth, development, long-term egg-laying capacity, and longevity of the queen. Thus, royal jelly has evolved as a complete diet for the queen. It contains nutrients as follows:

- Glucose and other simple sugars as carbon and energy sources
- Storage proteins with a balanced amino acid composition supporting new protein synthesis
- Trace minerals
- Medium-chain fatty acids (C-8 to C-10), presumably as energy boosters
- Hydroxy fatty acids, especially chain length C-8 to C-10, thought to contribute antimicrobial activity

Lipids of royal jelly are composed mainly of eight to ten carbon acids, hydroxy acids, and diacids, which may be saturated, unsaturated, linear, or branched. They include hexanoic acid, octanoic acid, E-oct-2-enoic acid, 8-hydroxyoctanoic acid, 3- and 10-hydroxydecanoic acid, and 3,10-dihydroxyoctanoic acid. A dramatic rise in 10-hydroxydecanoic acid levels in summer is believed to coincide with the need for increased antimicrobial activity during this season. Whereas workers are responsible for maintaining general sanitary conditions in the queen's nursery, royal jelly seems ideally suited as a substrate for preventing spoilage by bacteria and fungi. High levels of medium-chain fatty acids generating a low pH, along with the energy uncoupling properties of these chains, seem to act as a preservative against spoilage of royal jelly.

The structure of hydroxy fatty acids features a hydroxyl moiety (-OH) attached to the opposite end of the fatty acid chain from the carboxyl group. These molecules have three distinct chemical properties that might contribute to their bioactivity—interior region of six to eight saturated carbon atoms contributing lipid-loving properties, uncharged terminal hydroxyl groups with both lipophilic and lipophobic properties, and a terminal carboxyl group whose lipophilic/lipophobic properties depend on pH. When protonated, these acids are expected to freely shuttle protons across the membranes of would-be spoilage organisms, in essence uncoupling their proton gradients.

In addition to hydroxy fatty acids, saturated fatty acids, including hexanoic and octanoic acids, are present, perhaps as readily metabolizable energy sources. Recent data support a possible role for hydroxy fatty acids as signaling molecules activating TRPA1 (transient receptor potential cation channel A1) (Terada et al., 2011).

5.3   HYDROXY FATTY ACIDS ACTING AS SIGNALING MOLECULES MIGHT BE LINKED TO LONGEVITY OF QUEENS

TRP receptor proteins are a family of receptors embedded in sensory neurons for the purpose of detecting and amplifying a diverse and sometimes bizarre set of environmental signals. These signals include:

- Noxious chemicals
- Pleasurable chemicals (e.g., caffeine)
- Heat
- Cold
- Infrared
- Mechanical stretching

A vampire bat uses infrared-detecting TRP receptors located on sensory neurons on its face to target, in the dark, a landing spot where warm blood of a victim runs near the surface. A rattlesnake uses TRP receptors to accurately strike its victim even in complete darkness. TRP receptors are highly sensitive and tunable. Caffeine closes the Ca^{++} channel of TRPA1 receptors in mice, while causing the channel to open in humans. A family in Colombia has inherited a rare mutation in TRP that continuously signals excruciating pain.

Recently, a novel class of fatty acids present in royal jelly has been found to activate TRPA1 in vitro (Terada et al., 2011). This is a clue that dietary hydroxy fatty acids might modulate important metabolic processes, perhaps impacting longevity. This receptor is known to regulate energy homeostasis and thermogenesis, both likely important aspects of bioenergetics that might help modulate longevity. The nature of binding of hydroxy fatty acids to TRPA1 is of interest because these chains have dual chemical properties, being membrane soluble due to the hydroxy end of the molecule, in contrast to the carboxyl end favoring solubility in the aqueous phase at physiological pH. This chemistry is mentioned because TRPA1 might be activated by direct TRPA1-hydroxy chain binding or indirectly via effects of hydroxy fatty acids first interacting with membrane lipids. Bulk amounts of hydroxy fatty acids are fed to the queen, levels that seem unusually high compared to other classes of conventional signaling molecules. Thus, it would not be surprising that some chains of hydroxy acids are inserted into membranes, altering their physiological properties.

5.4   POLYUNSATURATED FATTY ACIDS MISSING FROM ROYAL JELLY MAY BE A SECRET TO LONGEVITY OF QUEENS

Haddad and colleagues (2007) suggest that the absence of long-chain fatty acids, especially polyunsaturated fatty acids, in royal jelly of bees enables longevity of queens. Data supporting this view come from their comprehensive analysis of membrane phospholipid composition of larvae, workers, and queens. The main result is that cell membranes of both young and old honeybee queens are highly monounsaturated with very low content of polyunsaturated chains. Newly emerged workers have a similar membrane fatty acid composition compared to queens. However, within the first week of life confined to the hive, workers increase their membrane content of polyunsaturated chains while decreasing the levels of monounsaturated fatty acids. Haddad and coworkers (2007) propose that the spike in polyunsaturated chains in membranes of workers is the result of pollen consumption.

Polyunsaturated chains seem to be a prerequisite for maturation and function of muscles enabling flight (Magwere et al., 2006). Thus, a positive feedback loop is established between pollen consumption and flight, enabling workers to acquire more pollen and nectar for the hive. Haddad and colleagues (2007) hypothesize that membranes of foraging workers become more susceptible to lipid peroxidation in the first week of life in the hive. Thus, membrane adaptation to flight can be considered to shorten life span. Seasonal temperature changes may also modulate longevity, with summer (warm weather) bees being short-lived compared to longer-lived winter (cool weather) bees. A membrane peroxidation model of longevity seems to account for these data (Figure 5.1), which is consistent with membrane unsaturated fatty acid composition being a determinant of maximum life span in bees. Recent data suggest a similar trend for queens of carpenter ants.

5.5 DIETARY MANIPULATION OF PROTEIN LEVELS ALSO HAS DRAMATIC EFFECTS ON LONGEVITY OF ANTS

The importance of diet in modulating aging in social insects has recently been demonstrated once again in the case of worker ants (Dussutour and Simpson, 2011). These researchers show that ant workers die young and colonies collapse when fed a high-protein diet. Perhaps the most striking finding is that feeding of workers with a high-protein diet only for a single day decreases worker longevity, causing a 20 percent reduction of the population size of the colony. The molecular basis of how a high-protein diet impacts worker longevity is not known, but these data suggest that nutritional homeostasis in ants plays a key role in enabling longevity. That is, major deviations from the normal diet, in this case involving protein or amino acid balance, can have negative effects on longevity of workers. One general explanation is that the diet of ants in nature is composed largely of carbohydrate-rich foods that are relatively low in protein. Nitrogen limitation often seems to prevail in areas where ants forage for food, with the result that high-protein food might be a rarity. Long-term adaptation of ants to a low-protein diet might result in a tendency to gorge when a high-protein meal becomes available, with potential effects on longevity of workers. It is of great interest to understand not only how a high-protein diet is amplified to decrease longevity of workers of ants, but also whether other social insects, including bees and termites, display a similar response to a high-protein diet. Interestingly, termites are among a few animals that support nitrogen-fixing bacteria in their gut, a testament that fixed nitrogen is likely a limiting nutrient for termites.

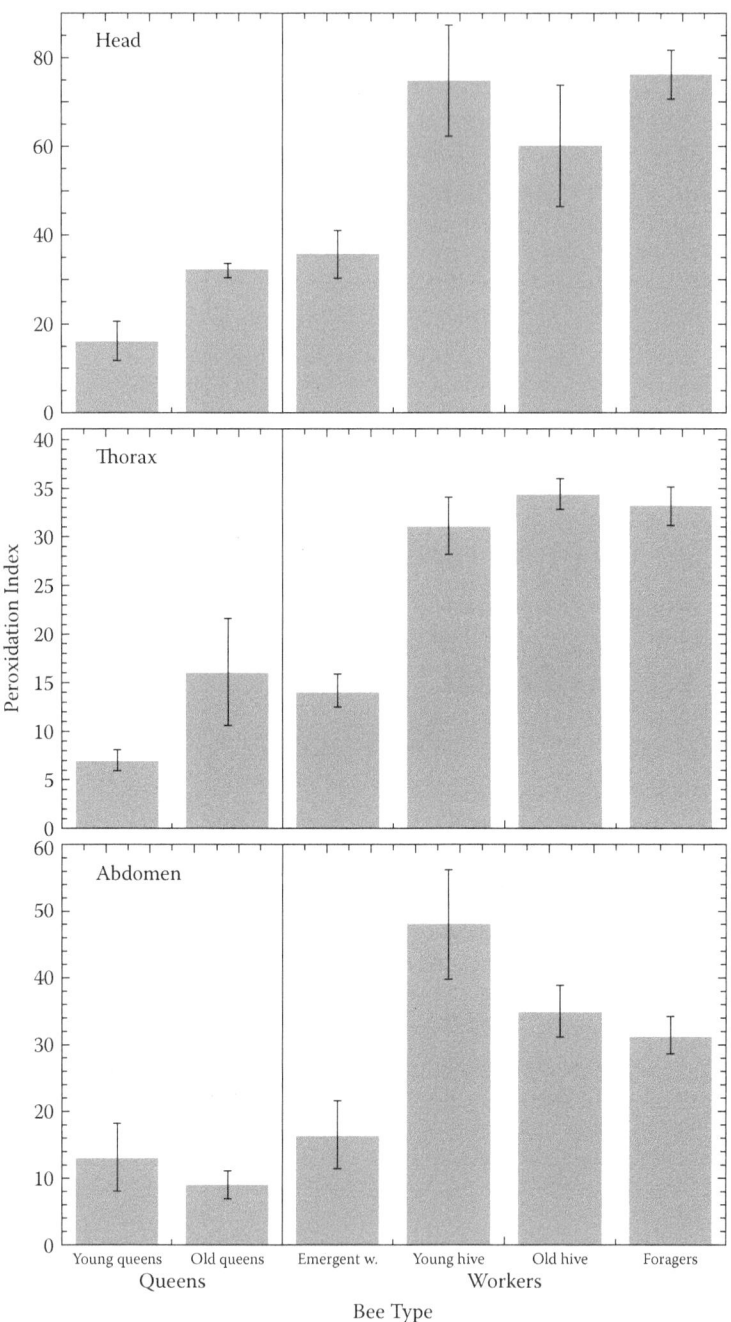

FIGURE 5.1 Dramatic lowering of peroxidation index of phospholipids of queen bees versus workers. These data support the membrane peroxidation theory of aging discussed in Chapter 6. See text for details. (Reprinted from Haddad et al., *Exp. Gerontol.* 42:601–9, 2007. Copyright © 2007. With permission from Elsevier.)

5.6   LESSONS FROM *DROSOPHILA*

Drosophila is a well-established model organism for research on aging with applications for humans. As background, most terrestrial insects, including flies, bees, and butterflies, produce no detectable DHA and only traces of EPA. Polyunsaturated chains containing fewer double bonds replace these highly unsaturated chains in membranes in this important group of insects.

Life span in *Drosophila* is modulated by various environmental signals. The following environments and conditions are known to extend life span in *Drosophila*:

- Cool temperatures
- Dietary restriction
- Mild stresses, including heat shock and cold stress
- Hypergravity
- Low levels of radiation
- Reduced flight activity

The strongest environmental stimulus that affects life span in *Drosophila* is environmental temperature, with flies cultured at 18°C living twice as long as those grown at 25°C. Temperature also modulates heart rate from 271 beats per minute at 22°C compared to 511 beats per minute at 38°C. Shifts in gross membrane fatty acid composition expected to accompany such environmentally mediated extension of life span have not been documented for *Drosophila*. However, studies of membrane biochemistry in *Drosophila* have led to a novel working model.

This model links environmental stimulus, membrane fatty acid unsaturation, and life span and is based on data from the pioneering research done by Schlame and colleagues on the molecular biology of cardiolipin (CL) synthesis in *Drosophila* (Malhotra et al., 2009; Schlame et al., 2005; Xu et al., 2009). These researchers found that the prominent species of CL in mitochondria of *Drosophila* is $(18:2)_4$-CL (Schlame et al., 2005). They point out that even though 18:2 is a minor fatty acid in *Drosophila* membrane phospholipids, an enigmatic mechanism triggered by environmental signals has evolved to generate a CL molecular species with up to four 18:2 chains. The specificity for CL molecular species did not reside in *Drosophila* taffazin, an enzyme that catalyzes synthesis of $(18:2)_4$-CL. Rather, unknown environmental signals are proposed to modulate synthesis of molecular species, including $(18:2)_4$-CL (Malhotra et al., 2009). These data suggest that the environment somehow directs the remodeling of mitochondrial CL in *Drosophila*, a mechanism that can roughly double the number of double bonds in CL from about four to eight, or vice versa, without the need for large changes in bulk membrane fatty acid composition. From a bioenergetic perspective, remodeling of molecular species of CL in *Drosophila* is expected to modulate major shifts in energy production, peroxidation rates, and other crucial physiologies of mitochondria.

These data are consistent with the mitochondrial inner membrane governing aging in *Drosophila* and are dependent on the following assumptions:

- Temperature is one of the signals modulating unsaturation levels of molecular species of cardiolipin.
- Cool temperature (17°C) increases levels of cardiolipin polyunsaturation (e.g., $(18:2)_4$-CL) but does not appreciably alter gross levels of unsaturated fatty acids compared to growth at 25°C.
- Cool temperatures are proposed to decrease rates of membrane peroxidation and approximately double longevity according to the membrane pacemaker model.

There is another aspect of aging in *Drosophila* that might be explained by the CL homeostasis. The heart organ of *Drosophila*, which pumps hemolymph through the body, is considered to be perhaps the hardest-working organ with the hardest-working mitochondria in *Drosophila*. It is known that *Drosophila* is subject to serious "heart disease," raising the question of whether the heart itself might be a critical single-organ pacemaker for aging in this animal (Ocorr et al., 2007; Vogler et al., 2009). The level of CL unsaturation might reach a peak in mitochondria of cardiac cells of *Drosophila*. That is, the mitochondrial membranes of heart cells in *Drosophila* might be enriched with $(18:2)_4$-CL or $(16:2)_4$-CL, creating unusually high levels of oxidative stress in the heart, in essence predisposing this vital organ to catastrophic oxidative damage.

5.7   DFOXO IN *DROSOPHILA*

There is increasing interest in the roles played by the universal transcription factor dFOXO in aging in *Drosophila*. In 2004 it was found that dFOXO controls life span and regulates insulin signaling in the brain and the fat body (Hwangbo et al., 2004). These researchers report the finding that dFOXO regulates *Drosophila melanogaster* aging at the whole animal level when activated in the adult pericerebral fat body. These data suggest the existence of a communication system linking the fat body with other organs. Also, in 2004 it was reported that insulin regulates heart function in aging fruit flies (Wessells et al., 2004). In this study dFOXO was found to influence age-dependent heart physiology and senescence directly, in addition to its systemic effect on life span. In 2007 dFOXO was shown to modulate a large number of genes involved in nutrition of *Drosophila* (Gershman et al., 2007). A total of 995 nutrient-responsive genes are regulated by dFOXO. In 2009 dFOXO was found to regulate organism size and stress resistance through an adenylate cyclase system (Mattila et al., 2009), and in 2010 dFOXO was shown to regulate organism-wide muscle homeostasis or proteostasis (Demontis and Perrimon, 2010). In this latter study decreased muscle aging was found to extend life span. Thus, dFOXO has been found to modulate numerous biochemical processes, including resistance to pathogens (Becker et al., 2010), with the power to govern life span in *Drosophila* (see Chapter 19 for discussion of FOXO in humans).

5.8   SUMMARY

Insects provide striking examples of how life span can be extended by dietary, thermal, and genetic manipulation. These data support the view that life spans of different animals are honed by Darwinian selection. An important lesson from studies of aging in insects concerns the remarkable extension of life span achievable by dietary and environmental factors. A second lesson highlights the importance of the linkage between membrane fatty acid unsaturation and longevity. It is especially interesting that dietary manipulation of levels of unsaturated membrane fatty acids appears to govern aging. Above all, these data support the view that longevity is a highly evolvable trait supporting the long-term goal of eventually extending disease-free human longevity.

REFERENCES

Becker, T., G. Loch, M. Beyer, et al. 2010. FOXO-dependent regulation of innate immune homeostasis. *Nature* 463:369–73.

Demontis, F., and N. Perrimon. 2010. FOXO/4E-BP signaling in *Drosophila* muscles regulates organism-wide proteostasis during aging. *Cell* 143:813–25.

Dussutour, A., and S. J. Simpson. 2011. Ant workers die young and colonies collapse when fed a high-protein diet. *Proc. Biol. Sci.* 279:2402–8.

Gershman, B., O. Puig, L. Hang, et al. 2007. High-resolution dynamics of the transcriptional response to nutrition in *Drosophila*: a key role for dFOXO. *Physiol. Genomics* 29:24–34.

Haddad, L. S., L. Kelbert, and A. J. Hulbert. 2007. Extended longevity of queen honey bees compared to workers is associated with peroxidation-resistant membranes. *Exp. Gerontol.* 42:601–9.

Honeybee Genome Sequencing Consortium. 2006. Insights into social insects from the genome of the honeybee *Apis mellifera*. *Nature* 443:931–49.

Hwangbo, D. S., B. Gershman, M. P. Tu, et al. 2004. *Drosophila* dFOXO controls lifespan and regulates insulin signalling in brain and fat body. *Nature* 429:562–66.

Kamakura, M. 2011. Royalactin induces queen differentiation in honeybees. *Nature* 473:478–83.

Keller, L., and M. Genoud. 1997. Extraordinary lifespans in ants: a test of evolutionary theories of ageing. *Nature* 389:958–60.

Keller, L., and S. Jemielity. 2006. Social insects as a model to study the molecular basis of ageing. *Exp. Gerontol.* 41:553–56.

Lercker, G., P. Capella, L. S. Conte, et al. 1981. Components of royal jelly. I. Identification of the organic acids. *Lipids* 16:912–19.

Magwere, T., R. Pamplona, S. Miwa, et al. 2006. Flight activity, mortality rates, and lipoxidative damage in *Drosophila*. *J. Gerontol. A Biol. Sci. Med. Sci.* 61:136–45.

Malhotra, A., Y. Xu, M. Ren, et al. 2009. Formation of molecular species of mitochondrial cardiolipin. 1. A novel transacylation mechanism to shuttle fatty acids between sn-1 and sn-2 positions of multiple phospholipid species. *Biochim. Biophys. Acta* 1791:314–20.

Mattila, J., A. Bremer, L. Ahonen, et al. 2009. *Drosophila* FoxO regulates organism size and stress resistance through an adenylate cyclase. *Mol. Cell. Biol.* 29:5357–65.

Melliou, E., and I. Chinou. 2005. Chemistry and bioactivity of royal jelly from Greece. *J. Agric. Food Chem.* 53:8987–92.

Ocorr, K., L. Perrin, H. Y. Lim, et al. 2007. Genetic control of heart function and aging in *Drosophila*. *Trends Cardiovasc. Med.* 17:177–82.

Pamilo, P. 1991. Life span of queens in the ant *Formica exsecta*. *Ins. Soc.* 38:111–19.

Schlame, M., M. Ren, Y. Xu, et al. 2005. Molecular symmetry in mitochondrial cardiolipins. *Chem. Phys. Lipids* 138:38–49.

Schneider, S. A., C. Schrader, A. E. Wagner, et al. 2011. Stress resistance and longevity are not directly linked to levels of enzymatic antioxidants in the ponerine ant Harpegnathos saltator. *PLoS One* 6:e14601.

Schrempf, A., S. Cremer, and J. Heinze. 2011. Social influence on age and reproduction: reduced lifespan and fecundity in multi-queen ant colonies. *J. Evol. Biol.* 24:1455–61.

Smith, C. R., C. D. Smith, H. M. Robertson, et al. 2011a. Draft genome of the red harvester ant *Pogonomyrmex barbatus*. *Proc. Natl. Acad. Sci. USA* 108:5667–72.

Smith, C. D., A. Zimin, C. Holt, et al. 2011b. Draft genome of the globally widespread and invasive Argentine ant (*Linepithema humile*). *Proc. Natl. Acad. Sci. USA* 108:5673–78.

Terada, Y., M. Narukawa, and T. Watanabe. 2011. Specific hydroxy fatty acids in royal jelly activate TRPA1. *J. Agric. Food Chem.* 59:2627–35.

Vogler, G., R. Bodmer, and T. Akasaka. 2009. A *Drosophila* model for congenital heart disease. *Drug Discov. Today Dis. Models* 6:47–54.

Wessells, R. J., E. Fitzgerald, J. R. Cypser, et al. 2004. Insulin regulation of heart function in aging fruit flies. *Nat. Genet.* 36:1275–81.

Wurm, Y., J. Wang, O. Riba-Grognuz, et al. 2011. The genome of the fire ant *Solenopsis invicta*. *Proc. Natl. Acad. Sci. USA* 108:5679–84.

Xu, Y., S. Zhang, A. Malhotra, et al. 2009. Characterization of tafazzin splice variants from humans and fruit flies. *J. Biol. Chem.* 284:29230–39.

6 Membrane Peroxidation Hypothesis Helps Explain Longevity in Birds, Rodents, and Whales

Bumblebee-sized hummingbirds live about five years. Pigeons can live thirty years, about seven times longer compared to a similar-sized rat (Montgomery et al., 2011). Naked mole rats from Africa can live thirty years or more compared to the three- to four-year life span of a similar-sized mouse. Large baleen whales might be the longest-lived mammals, with life spans estimated to be as long as two centuries. We have selected five animals of varying life spans to evaluate the membrane poly-unsaturation or peroxidation hypothesis of longevity (Figure 6.1) (Hulbert et al., 2007; Hulbert, 2008; Montgomery et al., 2011).

6.1 EXTREME FLIGHT OF HUMMINGBIRDS IS DEPENDENT ON HIGHLY UNSATURATED MITOCHONDRIAL MEMBRANES THAT MIGHT DICTATE THEIR SHORT LIFE SPAN

At least eight different physiological adaptations are necessary to enable extreme flight of hummingbirds, whose wing beats can reach greater than 100 per second during mating rituals and 50 to 80 per second during hovering. These adaptations include:

- Powerful heart and circulation system
- Unusually massive breast muscle
- Near maximum numbers of mitochondria in each breast muscle cell
- Double the normal levels of cristae in each mitochondrion
- Elevated body temperatures of around 40.6 to 41.7°C (compared to 37°C for humans)
- Enrichment of mitochondrial membranes with docosahexaenoic acid (DHA) (Infante et al., 2001)
- Possible synthesis of highly unsaturated molecular species of cardiolipin
- Many hours of deep nighttime torpor where body temperature can drop by 25°C and respiration rates can plummet fifty-fold compared to active flight

According to the rate of living theory of aging, the hardest-working mitochondria wear out most quickly, and consequently over a lifetime might undergo more rounds or cycles of division, including more replication cycles of their mtDNA (for historical

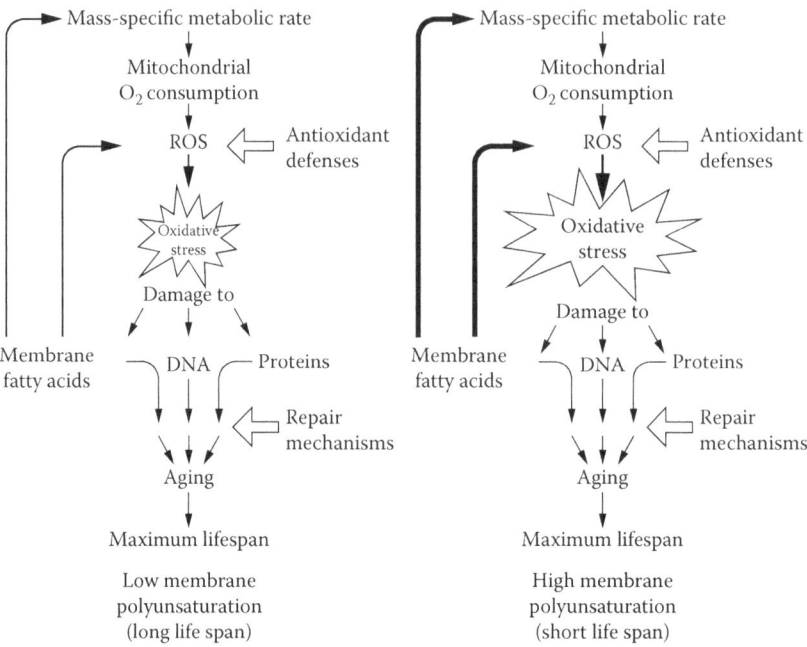

FIGURE 6.1 Membrane peroxidation hypothesis of aging. The left part of this diagram applies to oxidative stress in humans and other long-lived mammals, whose cellular membranes have a relatively low peroxidation index. The main point is that membranes with fewer double bonds enable longevity, and vice versa. The right side of the diagram applies to short-lived mammals such as mice with high content of DHA and polyunsaturated chains in membranes, and consequently great sensitivity to peroxidation, promoting aging. Recent molecular data support this general model with some revisions, as discussed in the text. (Reprinted from Hulbert, *Comp. Biochem. Physiol. A Mol. Integr. Physiol.* 150:196–203, 2008. Copyright © 2008. With permission from Elsevier.)

perspective on the rate of living theory, see Speakman, 2005; Speakman et al., 2002). According to the revised mtDNA theory of aging discussed in Chapter 9, each round of replication exposes mtDNA to natural errors or mutations that occur and accumulate as new strands of mtDNA are being laid down. Though the chance of a mutation is small at each round of mtDNA replication, this probability increases greatly when considered over the lifetime of the organism.

Mitochondria in hummingbird flight muscle have evolved to accommodate extreme flight, which in turn depends on extreme bioenergetics along with specialized membranes to support hyperrespiration of these mitochondria. We suggest that hummingbirds evolved their unique mitochondrial membrane unsaturation levels to keep up with energy demands for powering extreme flight. Thus, during flight these mitochondria are considered to operate in an energy stressed state. We predict that mitochondria for extreme flight and any mitochondria forced to operate near maximum capacity will wear out more quickly and require more cycles of division compared to mitochondria operating well below maximum horsepower. The point is that each division of mitochondria increases the probability for the accumulation of mutations in mtDNA—mutations that are proposed to eventually diminish cellular energy supplies and force the cell to a critical minimum threshold needed to sustain cellular functions; below this threshold cells undergo apoptosis. This concept is essentially a molecular restatement of the rate of living theory, incorporating recent advances in the molecular biology of aging.

Hummingbirds must feed voraciously and frequently to stay alive. Thus, any serious reduction of energy efficiency of flight muscle mitochondria caused by aging might quickly end the life of these birds. In the world of ecology it is difficult to find a better example of the consequences of age-dependent decline in mitochondrial energy supply and its corresponding effects on the life span of an animal.

Unfortunately, evolution of highly unsaturated fatty acid (HUFA)-enriched membranes—the adaptation that enables their amazing wing beats—is predicted to greatly shorten their life span (Hulbert et al., 2007). Life span among hummingbirds has a lower range of four to five years and an upper range of about twelve. A medium-sized hummingbird (broad-billed) banded and released in the wild holds the current longevity record of twelve years. Broad-billed hummingbirds have adapted to high elevations, and their migratory patterns take them to similar elevations and temperature patterns throughout their relatively short north-south seasonal migration route.

Hummingbirds adapted to life at high elevations display decreased rates of mutations in their mtDNA (Bleiweiss, 1998; this paper lists references of earlier work on the effect of the environment on mutation frequencies). These data suggest that the high-elevation environment directly influences rates of mtDNA mutations, perhaps prolonging life span. One scenario to explain the contribution of elevation to rates of mtDNA mutations is that at high elevation, nighttime temperatures almost always plummet. This would allow high-elevation hummingbirds to dependably harness extreme flight to acquire nectar during mild daytime temperatures while spending a large part of their life in a state of deep torpor brought on by consistently low nighttime temperatures. Note that hummingbirds in deep torpor require about twenty minutes to awaken. Body temperature often drops dramatically from 40.6°C to as low as 15.6°C. Also note that the torpid state caused by rapidly dropping temperatures characteristic of the Sonoran ecosystem drastically lowers respiration rates and energy consumption. Thus, the many hours each day spent in deep nocturnal torpor are expected to lower rates of oxidation of DHA, saving energy and in turn lowering spontaneous rates of mtDNA mutations. These data suggest that a lifetime of low nighttime temperatures inducing a deep, torpid state might significantly increase the life span not only of high-elevation hummingbirds, but also of hummingbirds in general. In essence, this example raises the possibility that hummingbirds have adapted to use low temperature as an avoidance mechanism to protect their DHA membranes and increase life span. Recall that *Drosophila* shares properties with hummingbirds in that cool temperatures of 17°C compared to 25°C can double their life span (Chapter 5). Obviously, further research is needed to establish a causal relationship between temperature and mtDNA mutation rates in hummingbirds.

6.2 SIMILAR BRAIN DHA LEVELS IN LONG-LIVED PIGEONS VERSUS SHORT-LIVED RATS SEEM TO DEFY POLYUNSATURATION THEORY OF AGING

According to the once popular theory of aging that states that maximum life span often increases roughly linearly with animal size, pigeons and rats, being of similar size, would be predicted to have similar life spans. Obviously, this is not the case, since pigeons live about seven times longer than rats (Montgomery et al., 2011). Montgomery and colleagues have recently revisited the pigeon-rat comparison, providing the most comprehensive membrane fatty acid data set yet to explain mechanisms behind the dramatic difference in maximum life span between these two similar-sized animals. The main conclusion is that the peroxidation index, a measure of rates of chemical oxidation of polyunsaturated membrane fatty acids, correctly predicts the seven-fold difference in life spans between these two animals. A number of other parameters often considered markers of aging were also tested during this study, but only membrane unsaturation levels seem to be of predictive value for judging longevity.

The data set on membrane fatty acid composition generated in the study by Montgomery et al. (2011) also provides fuel for thought regarding a number of other fundamental aspects of the molecular biology of polyunsaturated fatty acids in membranes linked to aging. First, the DHA theory of aging as originally envisioned by Gudbjarnason and discussed in this chapter might not have taken root had brain, kidney, or erythrocytes been chosen for fatty acid analysis instead of heart tissue. The point is that DHA levels in membranes of brain, kidney, and red cells of pigeons versus rats are similar in contrast to heart, liver, and muscles. Thus, DHA content of membranes of a specific organ such as the brain is not a universal predictor of life span. This point has been recognized by pioneers in this field for more than a decade and eventually led to the more universal peroxidation theory of aging (Figure 6.1).

The fatty acid data set generated by Montgomery et al. (2011) also illuminates the existence of dramatic tissue-specific differences in levels of polyunsaturated fatty acids such as 18:2 in addition to HUFAs, including arachidonic acid (ARA) and DHA. For example, membranes from rat brain tissue have six times more DHA than membranes from kidney tissue. This difference holds for pigeons as well. Different organs also display up to a ten-fold difference in 18:2 levels in membranes. We assume that these dramatic differences in species of unsaturated fatty acids are matched by equally important but unknown biochemical benefits. According to this scenario, mechanisms enabling specific fatty acids to be targeted to specific tissues, membranes, or organelles are predicted to play more important roles in membrane biochemistry than currently appreciated (see Section IV).

Montgomery and colleagues (2011) have asked the general questions of whether differences in longevity are driven mainly by systemwide failures or simply by failure of a key organ or system. It is also important to point out that specialized and critical biochemical demands placed on membrane systems such as vision have led to evolution of unique mechanisms to protect or recycle membranes. Thus, rhodopsin disks with extremely high peroxidation index values due to enrichment with DHA can be tolerated within rod cells of long-lived animals because of the specialized protection gained by continuous recycling of these membranes. It is also conceivable that an entire organ such as the brain might have evolved specialized layers of protection, which, in spite of a relatively high peroxidation index, protect this organ over the long term. This concept might help explain the fact that the peroxidation index of the brain tissue of long-lived pigeons seems to be greater than that of a short-lived rat (Montgomery et al., 2011).

There is an old concept in science that states that exceptions break long-accepted theories. Is the brain an exception that undermines the polyunsaturation theory of aging, or has the brain evolved specialized mechanisms providing long-term protection? This topic is covered in our recent book (Valentine and Valentine, 2013) and is an important area for future research (see Section IV).

6.3   NAKED MOLE RATS HAVE ONE-NINTH THE LEVEL OF DHA AS A MOUSE AND LIVE NINE TIMES LONGER

Naked mole rats (Figure 6.2) also seem to defy conventional theories of aging, a view supported by genomic analysis of this mouse-sized animal (Kim et al., 2011; Peterson et al., 2012a, 2012b). Some unusual properties of naked mole rats are as follows:

- Life span of more than thirty years compared to three to four years for a similar-sized mouse
- Negligible senescence
- No age-related increase in mortality
- Sexual activity through old age
- Resistance to spontaneous cancer and induced tumorigenesis
- Lack of age-dependent oxidative damage or accumulation of damaged proteins

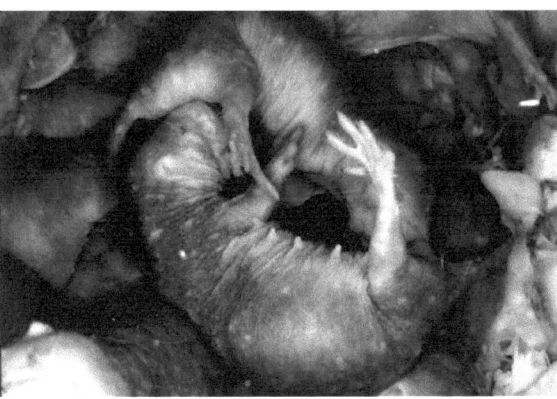

FIGURE 6.2   See color insert. Naked mole rat queen shown with one of her several consorts and other members of her colony. Like queens of social insects, including ants, a single naked mole rat queen gives birth to all individuals in her colony. This mouse-sized rodent outlives mice by almost ten times. A combination of diet, genes, and a unique underground environment is proposed to account for the cancer-free life and extraordinary longevity of this South African rodent. (From van der Horst et al., *BMC Evol. Biol.* 11:351, 2011.)

The lifestyle of naked mole rats (Bennett and Faulkes, 2000) is as extraordinary as their resistance to aging, cancer, and other stresses. Their life cycle involves:

- Residing in large underground colonies with a single breeding female or queen, who suppresses the sexual development of her junior females; only one to three males act as breeders
- Life in full darkness with eyes used only to distinguish between light and dark
- Living under subambient oxygen levels (about 8 percent O_2)
- Adaptation to high carbon dioxide concentrations in burrows (about 10 percent CO_2)
- Dependency on their underground environment for thermal control (i.e., normal body temperature between 30 and 32°C
- Evolutionary loss of certain pain signaling cascades common in mice and other mammals
- Don't require drinking water
- Recycling of nitrogen and other nutrients from feces
- Active microflora of intestinal symbionts that appear to help digest complex plant polysaccharides, which are major constituents of the naked mole rat diet
- Defective sperm similar to DHA knockout mutants of mice whose sperm are highly infertile, perhaps signaling a deficiency in DHA levels in naked mole rats (see Chapter 12)

Air in burrows is low in O_2 (about 8 percent) and high in CO_2 (greater than 10 percent) owing to many animals sharing a limited air supply and poor gas exchange through soil (Bennett and Faulkes, 2000). To adapt to the low O_2 conditions, naked mole rats have evolved several novel mechanisms, as follows:

- Altered hemoglobin oxygen affinity
- Reduced metabolic rate
- Slower rate of development
- Extreme hypoxia tolerance of brain function (Larson and Park, 2009)

In addition to adaptation to low O_2, the naked mole rat operates at a body temperature of 30 to 32°C and has poor thermoregulatory ability (Johansen et al., 1976). As discussed in Chapter 12, a lowering of temperature from 37°C (typical of humans) to between 30 and 32°C (typical of mole rats) is estimated to decrease membrane peroxidation rates by an order of magnitude.

Naked mole rats inhabit hot, arid regions in East Africa, where plants have adapted to this harsh environment by producing energy storage root structures, including tubers, corms, bulbs, or rhizomes (Bennett and Faulkes, 2000). These tissues provide carbon, protein, and other nutrients, in essence a staple food source for naked mole rats. Workers from a single colony can tunnel more than a mile to harvest these underground parts of plants, which they store in food caches in their burrows. In the wild and in captivity sweet potatoes are a favorite food for naked mole rats.

Interestingly, one of the largest colonies of naked mole rats ever recorded in the wild and composed of 295 individuals was located in a field of sweet potatoes in Africa. Usually colonies are much smaller, often limited by available food. A large tuber is often orders of magnitude greater in size than a 30 g mole rat and can sustain a colony for months. A large tuber is hollowed out, leaving the skin, which can "seed" growth of new shoots. Sweet potato used as food for the large colony mentioned above is composed of 93 percent carbohydrates, 6 percent protein, and only 1 percent fat. The fatty acid composition of the sweet potato tuber is about equally divided between saturated versus polyunsaturated chains and is markedly different compared to leaves of sweet potato. The leaves are strongly biased toward polyunsaturated chains of $18:2_{n-6}$ (linoleic) and $18:3_{n6}$ (linolenic), and the percentage of fat is higher. A relatively small level of starch is present.

Only the luckiest of naked mole rats dine on sweet potato, whereas most have evolved to survive on coarser and more indigestible food sources. Starch, starch-like storage forms, and other carbohydrates are the signature energy and carbon source stored in plant roots and essential for growth and survival of naked mole rats. This tissue also provides protein, some fats, and other key micronutrients. The emphasis here is on the quantity and quality of essential oils or precursors of essential fatty acids present in this diet.

Like a human or its close relative, the mouse, naked mole rats require omega-3 fatty acids and omega-6 fatty acids for building specialized membranes of neurons, sperm, and eyes. However, naked mole rats lack functional eyes, in contrast to other mammals, suggesting that the requirement for essential fatty acids and their precursors might be lower in naked mole rats. A clue behind this suggestion is that naked mole rat males produce highly degenerate sperm, similar to a mouse with a genetic block in DHA synthesis (reviewed by Valentine and Valentine, 2013).

The main points made here concern the biosynthesis of essential HUFAs by naked mole rats, as follows:

- The diet is devoid of preformed ARA, eicosapentaenoic acid (EPA), and DHA.
- Only a limited amount of α-linolenic acid ($18:3_{n-3}$) is present in the diet.
- Hydrogenation occurring in anaerobic regions of the gastrointestinal tract might further reduce levels of $18:3_{n-3}$, an essential precursor for omega-3 biosynthesis (see next section).

These data suggest that the diet itself and adaptations to this diet are mechanisms that drive down levels of HUFAs in naked mole rats, perhaps enabling their extreme longevity. We propose that numerous mechanisms besides low levels of HUFAs combine for increasing life span in this remarkable animal. For example, the extraordinarily low body temperature of this mammal might contribute to its longevity by lowering rates of membrane peroxidation.

We next look more closely at data generated during the mouse–naked mole rat comparative study reported by Hulbert et al. (2006). A comprehensive analysis of the DHA content of a variety of tissues of naked mole rats is reported and contrasted with simultaneous studies of mice. The startling finding is that naked mole rats have only about one-ninth the level of DHA as mice. For example, phospholipids from skeletal muscle, heart, kidney, and liver of naked mole rats contained an average 2.2 percent DHA (range 0.6 to 6.5 percent) versus 19.3 percent (range 11.7 to 26.2 percent) in mice. Even brain DHA content, which is constant among most mammalian species, is about 25 percent lower in naked mole rats than in mice. This drop in brain DHA might be linked to the fact that naked mole rats, like ants, are social animals with different workers or castes carrying out specific tasks. That is, brainpower can be shared, in essence diminishing the need for maximizing brainpower in any one individual. Once again, the most interesting result is that naked mole rats have only about 10 percent as much DHA and live roughly nine times longer than mice.

6.4   MEMBRANE POLYUNSATURATION MODEL PREDICTS WHALES AS THE LONGEST-LIVED MAMMALS

Gudbjarnason measured the levels of DHA in heart muscle tissue of various mammals ranging in size from small to large and from these data made two significant conclusions relevant to aging (Gudbjarnason et al., 1978; Gudbjarnason et al., 1989). The first is that DHA levels rise in lockstep with increasing heart rates, suggesting a direct relationship. Second, DHA levels in cardiac cells correctly predict life span, an inverse relationship. Gudbjarnason, who included whale heart tissue in his analysis, predicted that a large baleen whale might turn out to be the longest-lived mammal.

The claim that a giant baleen whale living in Alaska is the oldest living mammal is discussed next. The bowhead is an Arctic right whale with a large, bow-shaped head that is up to 40 percent of its body. In total bulk bowheads are second only to gigantic blue whales. The bowhead has made numerous adaptations for life in Arctic waters, where it resides year-round. The mouth of the bowhead, which gives rise to its name, is the largest and strongest of any mammal. The bowhead feeds by continuously swimming forward with its enormous baleen-covered mouth open. The baleen filter system allows grazing on copepods or zooplankton, which are the smallest, but most numerous, animals in the Arctic food web. The mouth has a large upturning lip on the lower jaw that reinforces baleen plates in its lower jaw and prevents buckling due to the strong water pressure exerted on the jaw as it feeds.

The bowhead is the only baleen whale that spends its entire life in extremely cold Arctic waters, with the Alaskan population migrating only as far south as the southwestern Bering Sea during the winter months. The group migrates northward in the spring, following openings in the pack ice, into the Chukchi and Beaufort seas hunting zooplankton. Like other right whales, bowheads swim slowly and, due to the often greater than 1 ft thickness of blubber surrounding their bodies, are extremely cold tolerant. Their massive blubber content acts as a large reserve of stored energy, preventing starvation when not feeding, and providing buoyancy. During the whaling era this buoyancy was a curse since diving as an escape mechanism to avoid whalers is difficult, and once harpooned, bowheads remain on the surface. For whalers this was the "right whale."

Analyses of bowhead whale gastrointestinal content (Hazard and Lowry, 1984) show that at least in the Alaskan Beaufort Sea, zooplankton such as copepods and euphausiids are the principal prey. In samples of gastrointestinal contents from fifteen whales these planktonic organisms comprised 73 to 99 percent of the stomach contents by volume. These data are consistent with the unique mouth adaption of bowheads, which allows them to filter large amounts of zooplankton-rich waters to sustain their large bodies and energy reserves. Records show that other prey such as benthic organisms are also taken. Zooplankton, however, seem to be the main food source for bowheads, especially in the spring. The often observed absence of any food in the stomachs of bowheads suggests that these whales depend on their enormous supply of blubber for energy when food is scarce, especially in winter months. Bowheads spend their entire lives in Arctic waters where much of the year is dark and not favorable for growth of their preferred prey. Thus, these whales can be considered to have adapted to a state of natural dietary restriction with intense feeding in spring and summer followed by long periods when food is scarce. Their giant storage of blubber makes this pattern possible and might favor a long life span.

Jeffrey Bada of the Scripps Institute of Oceanography is quoted as saying in a July 13, 2006, *National Geographic News* article that about five percent of bowhead whales are over 100 years of age, and in some cases, 160 to 180 years old (Roach, 2006). He points out that these are truly aged animals, perhaps the most aged on earth. These estimates are based on chemical analysis involving levels of amino acid isomerization seen in proteins in the nucleus of the eye lens (George et al., 1999). Lens proteins when first synthesized during early growth and development contain only l amino acid. Lens proteins, once made, remain permanently. This allows time for the chemical process of amino acid racemization (l → d amino acids) to occur, allowing Bada to use the ratio of l to d amino acids as a time clock for determining bowhead life span. These chemical studies, originally questioned by some researchers as not being biologically definitive, have since been reinforced by work of wildlife biologist Craig George of the Alaska Department of Wildlife Management. In 1990 George examined several whales taken during an annual hunt by Inupiat Eskimos, who have eaten whales as a major food source for centuries and are still permitted to harvest bowheads (George et al., 1999). George found stone harpoons embedded in whales, a method of hunting replaced by metal harpoons around 1860 to 1870. This finding raised the possibility that bowheads may reach 100 years of age. Similar findings of antique ivory spear points in whales killed in 1993, 1995, 1999, and 2007 triggered chemical studies by Bada as described above, searching for confirmation of bowhead ages. In May 2007 a bowhead caught off the Alaskan coast was discovered with the head of an explosive harpoon embedded deep under an old wound in its neck. Examination dated this arrow-shaped projectile manufactured in New Bedford, Massachusetts, to around 1890—a period when New Bedford was a whaling center. This suggests that this whale survived a hunt a century ago, placing its age at about 115 to 120 years. This value can be compared with the five whales dated by Bada's isomerization technique to be 91, 135, 159, 172, and the oldest, at 211 years old. The oldest might have been swimming off the Alaskan coast during the presidency of Thomas Jefferson. Japanese scientists have dated blue whales up to 110 years of age and a fin whale at 114 using a different method not applicable to bowheads.

George suggested that the bowhead's long life span might be due to its environment—cold water without abundant food available for long periods—forcing it to maintain a great body mass, an effective system for fat storage, an efficient insulation mechanism to keep warm, and often a starvation diet. The stress of living in Arctic waters may nurture the whale's pattern of slow growth and long life. Does the membrane pacemaker hypothesis explain the apparent long life span of bowheads?

Bowheads consume vast quantities of DHA/EPA present in zooplankton, with only a miniscule amount of total DHA/EPA from the diet needed to build and maintain specialized membranes, including those of the eyes and neurons. DHA/EPA content of blubber is reduced about 80 percent compared to their main food source (Reynolds et al., 2006). The reverse is true of monounsaturates, whose content rises from about 5 percent in the diet to about 30 percent in blubber. These data can be explained if the bowhead and its copepod prey team up to dramatically, metabolically alter unsaturation levels of dietary fatty acids (Christie, 1981; Budge et al., 2008). Data on dietary habits of bowheads suggest that the whale's digestive system, in conjunction with lipid dynamics and storage of lipids in its prey forms, has an ideal composition of fatty acids as gauged by the membrane peroxidation hypothesis of aging. Much of the responsible biochemistry has yet to be resolved. But one explanation is that whales, which evolved from ruminants (nearest living relatives are hippopotami), have retained powerful biohydrogenation reactions carried out by gut symbionts that convert excessive levels of dietary DHA and polyunsaturated fatty acids in the forestomach to more saturated forms. Hydrogenation of unsaturated fatty acids in the whale stomach prior to entering the bloodstream might be a win-win-win situation. The unsaturated fatty acids serve as alternative electron acceptors supporting higher rates of anaerobic respiration and robust growth of microbial symbionts, increasing caloric content of fatty acids due to reduction of double bonds, and perhaps ultimately increasing longevity. Note that bacterial biomass produced in the rumen is itself used as a source of food for ruminants. The meat of a bowhead is lean and has a fatty acid composition consistent with both premetabolism of dietary HUFAs and the operation of mechanisms targeting DHA to specific cells and membranes and away from most other cells, as in humans. We interpret the data of Gudbjarnason et al. (1978; Gudbjarnason, 1989) on fatty acid composition of whale heart tissue as evidence that DHA-enriched molecular species of cardiolipin are virtually absent from cardiac mitochondria and likely most other cells in the whale.

The hard shells of zooplankton consumed in great quantity by bowheads might also favor this whale's longevity. That is, chitin, the polymeric material forming the shell of zooplankton, is composed of a repeating unit of glucosamine, an energy-rich substrate released as glucose by symbionts in the rumen. Gut microbes are known to hydrolyze chitin and release large amounts of fermentable carbohydrates, a substrate that provides additional reducing power necessary to hydrogenate and restructure dietary fatty acids in the whale's forestomach, perhaps another biochemical process favoring longevity. Thus, it is proposed that an excess of dietary HUFAs is rapidly biohydrogenated prior to entering the bloodstream. In essence, we hypothesize that whales rapidly biohydrogenate dietary HUFAs in their rumen-like forestomach, enabling longevity.

6.5 SUMMARY

Data from research on aging of birds, rodents, and whales support the view that longevity is an evolvable trait geared to the life cycle of each kind of animal. The result is a remarkable level of diversity in life spans that differ by orders of magnitude in the case histories discussed here. At first glance this dramatic diversity in life spans seems to overshadow attempts to find unified mechanisms of aging shared by these animals. However, a closer look suggests that membrane peroxidation rates are a useful tool in predicting life span.

REFERENCES

Bennett, N. C., and C. G. Faulkes. 2000. *African mole-rats: ecology and eusociality*. Cambridge: Cambridge University Press.

Bleiweiss, R. 1998. Slow rate of molecular evolution in high elevation hummingbirds. *Proc. Natl. Acad. Sci. USA* 95:612–16.

Budge, S. M., A. M. Springer, S. J. Iverson, et al. 2008. Blubber fatty acid composition of bowhead whales, *Balaena mysticetus*: implications for diet assessment and ecosystem monitoring. *J. Exp. Mar. Biol. Ecol.* 359:40–46.

Christie, W. W. 1981. *Lipid metabolism in ruminant animals*. New York: Pergamon Press.

George, J. C., J. Bada, J. Zeh, L. Scott, et al. 1999. Age and growth estimates of bowhead whales (*Balaena mysticetus*) via aspartic acid racemization. *Can. J. Zool.* 77:571–80.

Gudbjarnason, S. 1989. Dynamics of n-3 and n-6 fatty acids in phospholipids of heart muscle. *J. Intern. Med. Suppl.* 731:117–28.

Gudbjarnason, S., B. Doell, G. Oskardottir, et al. 1978. Modification of cardiac phospholipids and catecholamine stress tolerance. In C. deDuve and O. Hayaishi (eds.), *Tocopherol, oxygen and biomembranes*. Amsterdam: Elsevier, pp. 297–310.

Hazard, K. F., and L. F. Lowry. 1984. Benthic prey in a bowhead whale from the northern Bering Sea. *Arctic* 37:166–68.

Hulbert, A. J. 2008. The links between membrane composition, metabolic rate and lifespan. *Comp. Biochem. Physiol. A Mol. Integr. Physiol.* 150:196–203.

Hulbert, A. J., S. C. Faulks, and R. Buffenstein. 2006. Oxidation-resistant membrane phospholipids can explain longevity differences among the longest-lived rodents and similarly-sized mice. *J. Gerontol. A Biol. Sci. Med. Sci.* 61:1009–18.

Hulbert, A. J., R. Pamplona, R. Buffenstein, et al. 2007. Life and death: metabolic rate, membrane composition, and life span of animals. *Physiol. Rev.* 87:1175–213.

Infante, J. P., R. C. Kirwan, and J. T. Brenna. 2001. High levels of docosahexaenoic acid (22:6n-3)-containing phospholipids in high-frequency contraction muscles of hummingbirds and rattlesnakes. *Comp. Biochem. Physiol. B Biochem. Mol. Biol.* 130:291–98.

Johansen, K., G. Lykkeboe, R. E. Weber, et al. 1976. Blood respiratory properties in the naked mole rat *Heterocephalus glaber*, a mammal of low body temperature. *Respir. Physiol.* 28:303–14.

Kim, E. B., X. Fang, A. A. Fushan, et al. 2011. Genome sequencing reveals insights into physiology and longevity of the naked mole rat. *Nature* 479:223–27.

Larson, J., and T. J. Park. 2009. Extreme hypoxia tolerance of naked mole-rat brain. *Neuroreport* 20:1634–37.

Montgomery, M. K., A. J. Hulbert, and W. A. Buttemer. 2011. The long life of birds: the rat-pigeon comparison revisited. *PLoS One* 6:e24138.

Peterson, B. L., J. Larson, R. Buffenstein, et al. 2012a. Blunted neuronal calcium response to hypoxia in naked mole-rat hippocampus. *PLoS One* 7:e31568.

Peterson, B. L., T. J. Park, and J. Larson. 2012b. Adult naked mole-rat brain retains the NMDA receptor subunit GluN2D associated with hypoxia tolerance in neonatal mammals. *Neurosci. Lett.* 506:342–45.

Reynolds III, J. E., D. L. Wetzel, and T. M. O'Hara. 2006. Human health implications of omega-3 and omega-6 fatty acids in blubber of the bowhead whale (*Balaena mysticetus*). *Arctic* 59:155–64.

Roach, J. 2006. Rare whales can live to nearly 200, eye tissue reveals. *National Geographic News.* http://news.nationalgeographic.com/news/2006/07/060713-whale-eyes.html.

Speakman, J. R. 2005. Body size, energy metabolism and life span. *J. Exp. Biol.* 208:1717–30.

Speakman, J. R., C. Selman, J. S. McLaren, et al. 2002. Living fast, dying when? The link between aging and energetics. *J. Nutr.* 132(6 Suppl 2):1583S–97S.

Valentine, R. C., and D. L. Valentine. 2013. *Neurons and the DHA principle.* Boca Raton, FL: Taylor and Francis Group.

van der Horst, G., L. Maree, S. H. Kotzé, and M. J. O'Riain. 2011. Sperm structure and motility in the eusocial naked mole-rat, *Heterocephalus glaber*: a case of degenerative orthogenesis in the absence of sperm competition? *BMC Evol. Biol.* 11:351.

7 Did Longevity Help Humans Become Super Humans?

The term *super ant* has been applied to this well-known social insect, which in spite of its relatively small brain has evolved an amazingly successful lifestyle (Chapter 5). One of the secrets to the success of ants that might apply to humans is that of specialization of tasks that maximize the efficiency of the colony. Nurse ants feed and clean the queen, maintaining a healthy environment in the nursery. Warriors fend off invaders while food gatherers feed the colony. Farmer ants grow edible fungi in the humid interior of the mound and morticians remove the dead. When the most specialized of all ants, the queen, dies prematurely, the colony often collapses. Thus, the queen's extreme longevity of up to twenty years (about twenty times longer than that of a worker) is of great selective advantage in the world of ants. The selective value of longevity for humans remains unknown, although numerous theories have been proposed.

7.1 CASPARI'S HYPOTHESIS OF THE RISE OF GRANDPARENTS

Recent archeological data suggest that humans old enough to become grandparents appeared relatively recently in the history of human evolution (Caspari and Lee, 2004; Caspari, 2011). One explanation is that human longevity was not of great selective advantage during millions of years of human evolution but became a powerful selective advantage in a relatively short time period. Prior to about 30,000 years ago, the average life span of a human appeared to be 30 or so years, leaving no time in the life cycle for the existence of significant numbers of grandparents compared to the present, when even great grandparents are common (Figure 7.1). According to this scenario human longevity had its golden era in terms of beneficial selective advantages mainly over the last 30,000 years. By this time our ancestors are considered to be modern humans closely resembling humans today. The specific selective advantages sufficient to usher in the era of grandparents starting thousands of years ago remain largely unknown.

It is interesting to speculate that longevity helped humans become super humans, perhaps like the case of super ants (Chapter 5). Caspari addresses this question emphasizing the parallel rise of grandparents with a quantum leap forward in cultural activity along with an increase in population size. We suggest that a more reliable food source, both nutritious and energy-rich, might have been an important factor in selection of grandparents (O'Connell et al., 1999; Trammell et al., 2008).

FIGURE 7.1   See color insert. Grandparents and great grandparents emerged late in human evolution. Recent archeological data suggest that prior to 30,000 years ago life spans were too short to achieve grandparenthood. Genetic, sociological, and environmental changes that might have enabled a spike in human longevity remain a mystery. From a genetic perspective, it is now known that mutations in one or a small number of genes or DNA regulatory elements are able to enhance longevity of animals, raising the possibility that longevity genes are important in the evolution of human life span. The photo shows four generations in the family of Carla Valentine, MD: Carla with daughters Stella (held), Isabella, and Sienna, and Carla's mother and grandmother. (Courtesy of David L. Valentine, PhD.)

This idea is consistent with concepts developed here and in our previous two books that human longevity is a selectable trait that comes with a high bioenergetic cost. In addition, the diet and energy efficiency of modern humans living 30,000 years ago had been honed by numerous cycles of dramatic changes in the environment, including glaciation.

7.2   FOSSIL TEETH SHOW LONGEVITY OCCURRED LATE IN HUMAN EVOLUTION

Analyses of the fossilized teeth of hundreds of individuals spanning 3 million years from Australopithecus—early Homo—Neanderthal to early modern European were used to determine life spans of human ancestors (Caspari, 2011). These data show that living long enough to reach grandparenthood became common relatively late in human evolution. Evaluation of the proportion of grandparent-aged adults relative to younger adults led to the striking conclusion that the ratio increased only slightly over most of human evolution. Then around 30,000 years ago longevity spiked sharply higher. Analysis of fossilized teeth of Neanderthals shows that for every ten young adults who died between the ages of fifteen and thirty, there were only four older adults who survived past age thirty. The age distribution of deaths of European

Upper Paleolithic individuals was remarkably different. For every ten young adults in this society there were twenty potential grandparents. The criterion for lack of grandparenthood is based on a reproductive cycle with the first child born to a mother fifteen years of age and reared to childbearing age at fifteen. At this point the mother of the first child had completed her life and died on average at around thirty. This life cycle leaves little room for grandparenthood or any of its benefits.

7.3 ANCIENT ARTIFACTS SHOW CULTURAL REVOLUTION AND POPULATION GROWTH COINCIDING WITH LONGEVITY

The rise of grandparents (increased longevity) coincides with a dramatic shift to more sophisticated weapons and art compared to the relatively simple technology of the preceding period. The data from artifacts of tools and weapons and art do not prove a causal relationship between longevity and an explosion in culture during this period, but they are consistent with an important linkage. There is also the question of the existence of a positive feedback loop operating between longevity and culture, allowing each to reinforce the other. Caspari (2011) addresses this topic in her article in *Scientific American* and cites data from other experts on how a positive feedback mechanism linking longevity and cultural dynamics might have enabled population growth.

We suggest that a reliable supply of food (O'Connell et al., 1999; Trammell et al., 2008) is a necessity to support both population growth and longevity based on the high bioenergetic cost of both phenomena. Therefore, any set of conditions, whether human inspired or environmentally mediated, that improved the food supply had the potential to trigger a cultural revolution and a trend of increased population growth such as occurred 30,000 years ago.

7.4 DOES BIOENERGETICS HELP EXPLAIN THE EVOLUTION OF GRANDPARENTS?

The bioenergetics of modern humans was honed during numerous cycles of climate change leading up to the last ice age about 30,000 years ago. Remarkably, the Neanderthals who dominated the European continent for 200,000 years became extinct about the time that longevity of modern humans was increasing. It is clear that changes in climate forced our modern human ancestors to adapt to climate-mediated changes in food supply. In essence, modern humans became increasingly sophisticated in acquiring, preserving, and storing food, which allowed them to survive when food was scarce. In addition, modern humans prior to the last ice age had a diverse diet compared to Neanderthals, the latter of which seemed to specialize as hunters of large ruminant animals. What does bioenergetics have to do with the rise of longevity? The answer, as we argue throughout this book, is that longevity is an energy-intensive process. Consequently, availability of food might have been an important factor in evolution of grandparents.

We suggest that more food energy was available during this era due to a more abundant and reliable food supply, setting the stage for genetic changes enabling longevity. Recent data suggest that relatively few genetic changes might have led to increased longevity of humans. As discussed in Chapter 19, much of the human genome encodes DNA regulatory elements representing a multitude of targets for mutations increasing longevity. That is, new genes and new functions are not needed, as longevity can be achieved by minor changes to regulatory networks.

7.5   EVOLUTION OF HUMAN LONGEVITY AS A GENETIC EVENT

It is not known what evolutionary or adaptive changes took place in our ancestors to bring about the surge in longevity starting about 30,000 years ago. However, we likely owe our current longevity to these events. By 30,000 years ago it is likely that our ancestors had evolved all six of the homologs of the FOXO transcription factors suggested to protect human cells against aging (see Chapter 19). Indeed FOXO3A, a genetic variant of FOXO3 found in centenarians living today, serves as a model to explain how genetic changes might have occurred, giving rise to a spike in human longevity. FOXO3A and other putative longevity genes are suggested to enable a healthy life for up to a century. FOXO3A is expressed in the brain as well as other cells, where it acts to protect cells against various stresses linked to aging. Only a tiny proportion of the current human population carries this mutation (i.e., FOXO3A), but it appears to have evolved multiple times in different regions of the world. For example, Japan has the highest ratio of centenarians, with 34.7 Okinawans per 100,000 inhabitants living to 100 years of age. Immigrants from Japan brought FOXO3A to Hawaii (Willcox et al., 2008). We propose that a single mutation in the FOXO system of cellular protection, or genetic change in other longevity genes, is powerful enough to have caused the spike in human longevity described by Caspari (2011). Note that DNA mutations changing patterns of expression of longevity genes are expected to occur more often than structural mutations.

The search for longevity genes in humans as described in Chapter 19 has led to other candidates besides FOXO that link mitochondria, energy, and longevity. As background and as discussed in Chapter 19, there are increasing data showing that mitochondrial bioenergetics and longevity are closely linked. Indeed, recent data suggest that longevity is energy-intensive in keeping with the notion that our ancestors needed more energy food to pay the additional energy cost for longevity. In a strange quirk of nature, mitochondria appear to bleed off a significant amount of the energy they produce to protect their membranes and other targets against oxidative stress. Human mitochondria have evolved five different enzymatic spigots or portals called uncoupling proteins (UCPs 1–5) whose purpose is to release energy from mitochondria as a protective mechanism. This release of mitochondrial energy is not wasted, just as release of floodwater from a dam prevents washing out the foundation. Indeed, the current picture is that energy consumed to prevent oxidative damage to mitochondrial membranes is a bargain in terms of a gain in longevity. Recent data suggest that one or a few mutational changes occurring in mitochondrial uncoupling proteins increase longevity (Rose et al., 2011; see Chapter 19). In spite of much publicity, it is fair to say that matching specific genes or DNA regulatory elements to longevity is in its infancy.

7.6 CONVERGENT EVOLUTION OF LONGEVITY

It is clear that longevity is a highly selectable trait that has evolved on numerous occasions in different animal species (convergent evolution). It is also likely that during the history of animal evolution there are cases in which selective pressure for longevity arose and then disappeared when risks of longevity matched or outweighed benefits. However, we are aware of little information on such reverse cycles of longevity. Selection of longevity has benefited other long-lived animals, and the case of whales introduced in Chapter 6 is covered in more detail here.

The great baleen whales have evolved longevity equivalent to or greater than humans (see Chapter 6). Whales evolved from ruminants, with the hippopotamus being their closest living relative. The fossil record for whales has been filled in and shows that whales and hippos shared a common terrestrial ancestor about 60 million years ago. However, evolution of the great baleen whales began in earnest about 30 million years ago. This period, which followed major geological shifts of continents and cooling of ocean temperature, created choice ecosystems such as Antarctica, providing fertile feeding grounds supporting the massive energy requirements of great whales. Baleen whales evolved a brain more than five times the size of that of humans, providing brainpower for survival, including navigating great distances to and from their seasonal feeding grounds. Over a period of 30 million years, six-fold longer than required for human evolution, whales evolved longevity to match their lifestyle. It is easy to see the benefits of longevity for baleen whales, who nurture their offspring for long periods and whose remarkable navigational skills take years of experience to perfect. In other words, whale elders pass on to their offspring and clan a selective advantage for their species: knowledge that is essential for their survival.

It is also easy to imagine based on our own experience possible benefits of grandparents for human evolution. It is clear that our ancestors 30,000 years ago, like the great whales, were hunters and gatherers. It is interesting to speculate that like ants and whales, humans evolved as super humans by harnessing the collective brainpower, wisdom, and experience stored in the brains of all individuals, especially the elder generation. We also suggest that in addition to the wisdom of age, our ancestors may have benefited from gifted individuals whose brains were rewired along the lines of autistic children today. Thus, autistic tendencies might have once been a blessing in contrast to the stigma associated with this brain disorder today.

A list of possible benefits of grandparenthood is compiled below:

- Longer-term fertility
- Material or economic resources
- Current social position and past social connection
- History related to survival (e.g., long-term weather patterns)
- What to eat
- What not to eat
- When to eat it
- How to preserve and store food
- Water resources

- Advice on procedures (e.g., weaving a basket, producing a sharp cutting edge on a tool or weapon, making warm clothes)
- Locations of hard-to-find necessities (e.g., stone for tools or weapons)
- Medicinal and health advice
- Instructions on behavior
- Concepts of teamwork (e.g., cooperative hunting)
- Teaching related to art, communication, life and religion
- How to nurture wild plants and animals to produce a more reliable food source (i.e., prior to domestication as needed for farming)
- Childcare releasing parents to gather more food
- Communal living
- Other specialization (i.e., unique skills dependent on years of experience)

Undoubtedly, experienced grandparents living today will be able to greatly expand this list.

7.7   PRESENT-DAY RISKS OF HUMAN LONGEVITY

We next switch gears from possible advantages of longevity in the past to highlight a cardinal risk of human longevity that is emerging today. As described in our previous book (Valentine and Valentine, 2013), it is well known that aging is the major risk factor for Alzheimer's and other forms of dementia now reaching epidemic proportions in an aging world population (Figure 7.2). Alzheimer's and aging share many common mechanisms. One scientific benefit of this tight linkage is that advances in understanding aging are applicable to Alzheimer's, and vice versa. Note

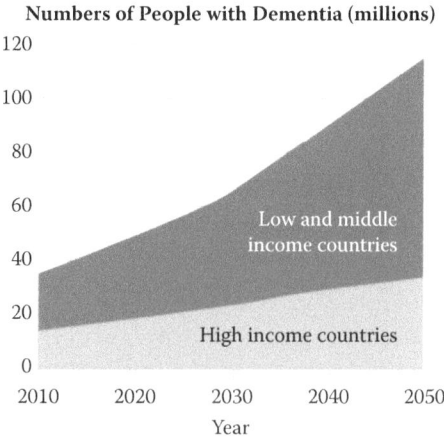

FIGURE 7.2   Alzheimer's Disease International, which monitors worldwide trends in neurodegenerative diseases, predicts a sharp increase in Alzheimer's disease as the world's population ages. Cases of neurodegenerative diseases, especially Alzheimer's disease, are expected to reach more than 100 million persons by 2050. (Used with kind permission from Alzheimer's Disease International, *World Alzheimer Report 2010*.)

that the Food and Drug Administration (FDA) does not recognize aging as a disease to be cured, in contrast to Alzheimer's. However, a paradigm shift is now occurring largely undermining this view (see Section I, Introduction). There is growing evidence that aging has earmarks of a genetic disease. The main point here is that we consider advances in knowledge of the causes of age-dependent diseases, including Alzheimer's and neurodegenerative diseases in general, to be immediately applicable to understanding human aging, and vice versa. Whereas the emphasis in the current book is on mechanisms of aging and longevity, we will on occasion switch back and forth between aging and neurodegeneration to make a point. The brain stands out as a single organ that determines life span, as discussed next.

It is estimated that by the year 2050 more than 100 million persons worldwide will experience early brain death, with the numbers of patients with symptoms of neurodegeneration doubling about every twenty years. One of the most troubling statistics is that 50 percent of people reaching age 85 years of age will display symptoms of neurodegeneration. It is unlikely that our ancestral population, with an average age of death of thirty years, faced an epidemic of Alzheimer's as we do today. The cause of this epidemic of brain diseases is clear, and it is aging itself. As mentioned above, a mutation or mutations driving up longevity from say 30 to 60 years might have sprung up and spread through modern humans starting about 30,000 years ago. As described in our previous book (Valentine and Valentine, 2013), neurons require specialized protection to ensure brain span. Today most of the human population, now at 7 billion persons, might be saddled with mechanisms of longevity that evolved to ensure a brain span of about sixty years. During the past century and due largely to modern medicine and reduction in rates of infant mortality, the average life span of humans in developed societies has spiked again and is now about eighty years. Thus, the age of longevity has ushered in an epidemic of neurodegeneration (Jackson and Prince, 2009).

7.8 REVISED HOLY GRAIL OF AGING

The holy grail of aging is changing. Events happen quickly. In 2008 it was reported that a longevity gene had been identified in centenarians (Willcox et al., 2008), creating considerable public interest in raising the bar for the human life span to 100 years (see Chapter 19). Is this a realistic goal? A reevaluation of long-term goals in the field of aging is now underway, driven by scientific advances balanced by ecological, financial, and social constraints. With the global population expected to reach 9 billion in the decades ahead, our infrastructure may not be able to keep up with the demand for feeding and healthcare of a growing population of centenarians. Also, a new social order would be necessary to accommodate so many aged persons. At this time the state of the art of the science of aging is such that it is not possible to predict if or when a century of life for every person is a realistic scientific goal. And the growing realization that the current era of longevity, with life expectancy now pegged at about 80 years in developed countries, has spawned a worldwide epidemic of Alzheimer's disease adds a sobering note for planners. These considerations dictate a revised holy grail of aging—a healthy or disease-free life through old age. Even this more modest goal represents a monumental challenge for the community of scientists working in the field of aging.

7.9   SUMMARY

Recent data suggest that grandparents were largely absent for most of human evolution up until about 30,000 years ago. By this time most other modern human traits were likely in place. According to a theory proposed by Caspari (2011), longevity of humans may have evolved rather quickly, starting about 30,000 years ago. This period coincides with population growth, major cultural development, and advanced tool and weapon-making skills, and might have set the stage for an advance in human longevity. We suggest that a stable food supply may have been linked to these events. The availability of more food for whatever reasons means that the human organism had more energy available—energy that might have been channeled toward longevity. There is increasing evidence that human longevity is paid with the currency of energy, a lifelong commitment for a significant fraction of total energy produced by mitochondria. If this concept is correct, then energy supply and longevity are interlinked. Because the energy supply for powering human longevity is so high, our ancestors prior to 30,000 years ago may not have benefited from longevity. This balance pitting benefits versus risks of longevity might be reversing itself in our current age of longevity; benefits of longevity may be balanced against great risks caused by an epidemic of neurodegenerative diseases such as Alzheimer's disease. On a positive note, there is evidence that genetic changes in longevity genes already seen in the human gene pool add protective power, enabling a healthy life through old age. Finally, the finding (discussed in Section III) that aging mammalian mitochondria decline in energy production may indicate that the relative bioenergetic cost of longevity increases as we age. To balance the putative high-energy cost of longevity, we suggest that grandparenthood provided an ecological benefit at some point over the past 30,000 years.

REFERENCES

Caspari, R. 2011. The evolution of grandparents. *Sci. Am.* 305:44–49.

Caspari, R., and S. H. Lee. 2004. Older age becomes common late in human evolution. *Proc. Natl. Acad. Sci. USA* 101:10895–900.

Jackson, J., and M. Prince (eds.). 2009, September 21. *World Alzheimer Report 2009— Executive Summary*. Alzheimer's Disease International.

O'Connell, J. F., K. Hawkes, and N. G. Blurton Jones. 1999. Grandmothering and the evolution of Homo erectus. *J. Hum. Evol.* 36:461–85.

Rose, G., P. Crocco, F. DeRango, et al. 2011. Further support to the uncoupling-to-survive theory: the genetic variation of human UCP genes is associated with longevity. *PLoS One* 6:e29650.

Trammel, J., J. O'Connell, S. Bush, et al. 2008. Exploring geophyte use in the Northern Great Basin: nutrient content, handling costs, effects of fire and tillage, and archaeological implications. http://www.anthro.utah.edu/PDFs/field_school/GBAC_08.6ppt.pdf.

Valentine, R. C., and D. L. Valentine. 2013. *Neurons and the DHA principle*. Boca Raton, FL: Taylor and Francis Group.

Willcox, B. J., T. A. Donlon, Q. He, et al. 2008. FOXO3A genotype is strongly associated with human longevity. *Proc. Natl. Acad. Sci. USA* 105:13987–92.

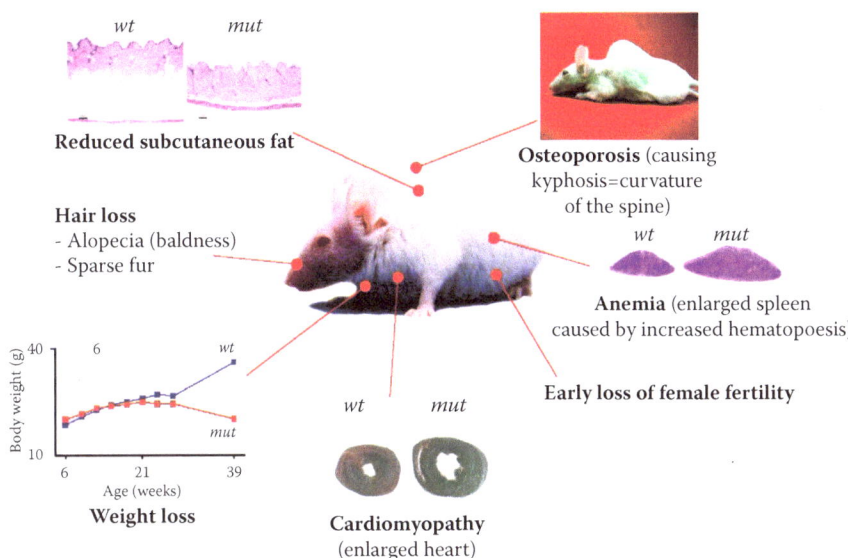

COLOR FIGURE 1.1 Aging phenotypes in mtDNA mutator mice. The mtDNA mutator mouse is genetically engineered to express a proofreading-deficient version of the mtDNA polymerase. This "mutator gene" leads to a three- to five-fold increase in somatic point mutations of mitochondrial DNA, an occurrence of a linear deleted mtDNA molecule, a progressive respiratory chain dysfunction, an expression of a variety of premature aging phenotypes, and a shortened life span. Humans harboring a naturally occurring mutator gene display numerous pathological phenotypes, as discussed in Chapters 8 and 9. (Reprinted from Edgar et al., *Cell Metab.* 10:131–38, 2009. Copyright © 2009. With permission from Elsevier.)

(a)

COLOR FIGURE 2.1 Paraquat catalyzes formation of reactive oxygen species (ROS), which can kill plants and humans and cause Parkinson's disease. (a) Photo of a corn leaf with lesions bleached by tiny droplets of paraquat drifting on the wind following spraying of a nearby weed field. (Photo used with permission from Dr. Kevin Bradley at the University of Missouri.)

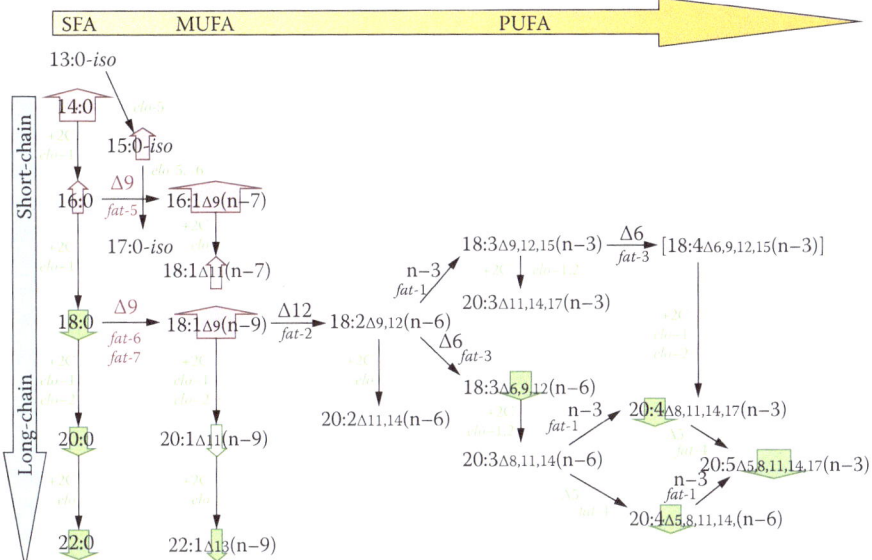

COLOR FIGURE 4.2 Regulatory patterns of biosynthesis of EPA and other unsaturated fatty acids linked to aging. See Shmookler Reis and colleagues (2011) for a detailed explanation of regulatory patterns of fatty acid synthesis in *C. elegans*. Three points are highlighted here: (1) Black arrows show the overall pathway of membrane fatty acid biosynthesis leading to EPA. (2) Enzyme activities (+2C for elongases; Δn, n-3, or n-6 for desaturases) and their genes are shown beside each arrow. (3) Bold (dark) font over the fatty acid classes indicates up- or down-regulation with increasing life span. The width of the arrow corresponds to the strength of the correlation to longevity. (Shmookler Reis et al., *Aging (Albany NY)* 3:125–47, 2011. Copyright © 2011 Shmookler Reis et al.)

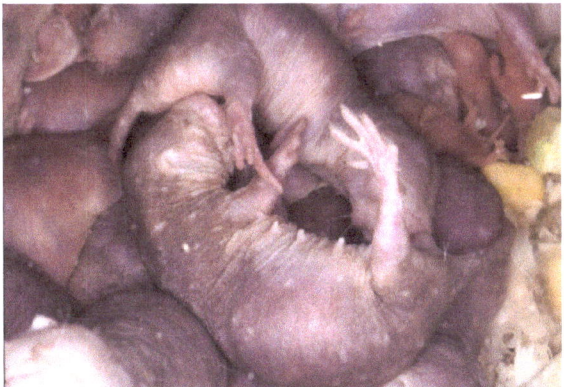

COLOR FIGURE 6.2 Naked mole rat queen shown with one of her several consorts and other members of her colony. Like queens of social insects, including ants, a single naked mole rat queen gives birth to all individuals in her colony. This mouse-sized rodent outlives mice by almost ten times. A combination of diet, genes, and a unique underground environment is proposed to account for the cancer-free life and extraordinary longevity of this South African rodent. (From van der Horst et al., *BMC Evol. Biol.* 11:351, 2011.)

COLOR FIGURE 7.1 Grandparents and great grandparents emerged late in human evolution. Recent archeological data suggest that prior to 30,000 years ago life spans were too short to achieve grandparenthood. Genetic, sociological, and environmental changes that might have enabled a spike in human longevity remain a mystery. From a genetic perspective, it is now known that mutations in one or a small number of genes or DNA regulatory elements are able to enhance longevity of animals, raising the possibility that longevity genes are important in the evolution of human life span. The photo shows four generations in the family of Carla Valentine, MD: Carla with daughters Stella (held), Isabella, and Sienna, and Carla's mother and grandmother. (Courtesy of David L. Valentine, PhD.)

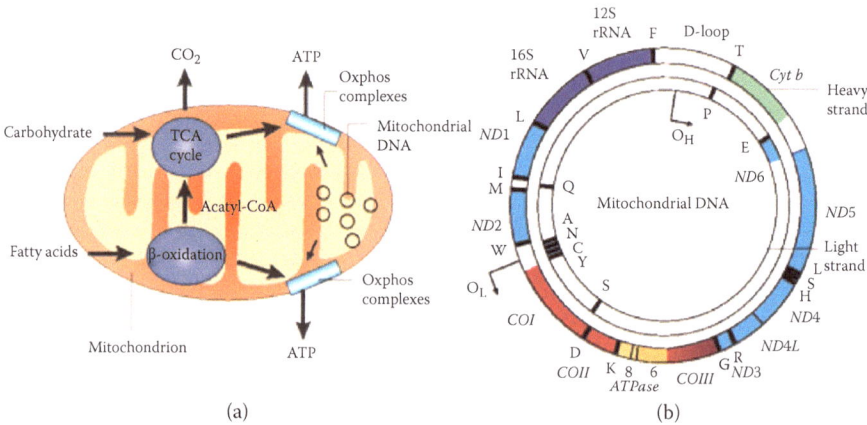

(a)   (b)

COLOR FIGURE 8.1 Human mitochondrial DNA (mtDNA) is a small, circular double helix encoding several essential energy-producing genes. The genes that encode the subunits of complex 1 (ND1–6 and ND4L) are shown in blue; the terminal complex, cytochrome c oxidase (COI-COIII), is shown in red; cytochrome b of complex 3 is shown in green; and the subunits of ATP synthase (ATPase 6 and 8) are shown in yellow. RNA genes are also listed (purple and black slashes). (From Taylor and Turnbull, *Nat. Rev. Genet.* 6:389–402, 2005. Copyright © 2005. Reprinted by permission from Macmillan Publishers Ltd.)

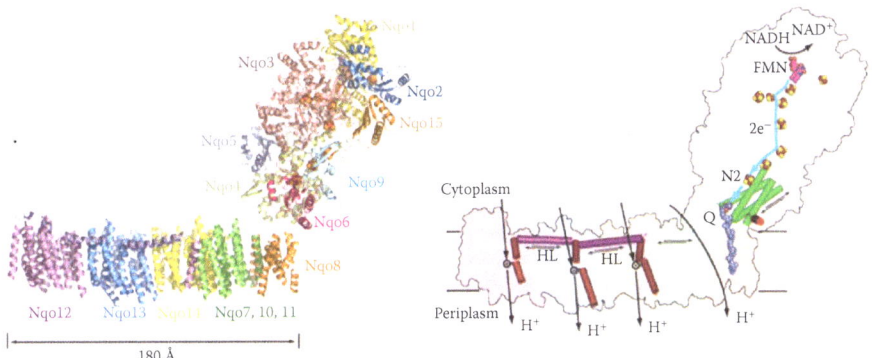

Cytoplasm

Periplasm

180 Å

NADH NAD⁺

FMN

2e⁻

N2

Q

HL HL

H⁺ H⁺ H⁺ H⁺

COLOR FIGURE 9.1 Structure of chair-shaped complex 1 showing numerous subunits (left) and topology of electron flow (right). Note that electrons flow from NADH as donor through iron-sulfur centers located in the long arm of the enzyme extending into the mitochondrial matrix. The proton-pumping portion of the enzyme is embedded in the membrane. High-energy electrons flowing from NADH to ubiquinone lose energy while energizing the efflux of protons. Complex 1, composed of forty-five different subunits, is by far the most complicated member of the electron transport chain and is a major target of mutations occurring in mtDNA. Complex 1 is the largest enzyme whose structure has been solved, and this achievement opens a new window for understanding membranes' contribution to longevity. (From Efremov et al., *Nature* 465:441–45, 2010. Copyright © 2010. Reprinted by permission from Macmillan Publishers Ltd.)

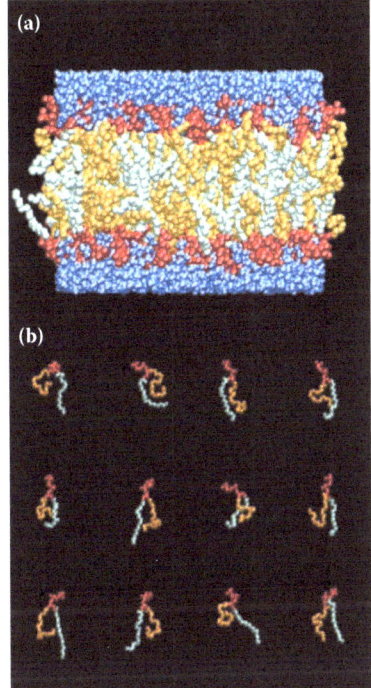

COLOR FIGURE 10.1 Extraordinary conformational dynamics of DHA. (a) DHA-enriched membrane. (b) Dynamic conformations of DHA phospholipids. (Images courtesy of Scott Feller, and generated by Matthew B. Roark, both of Wabash College.)

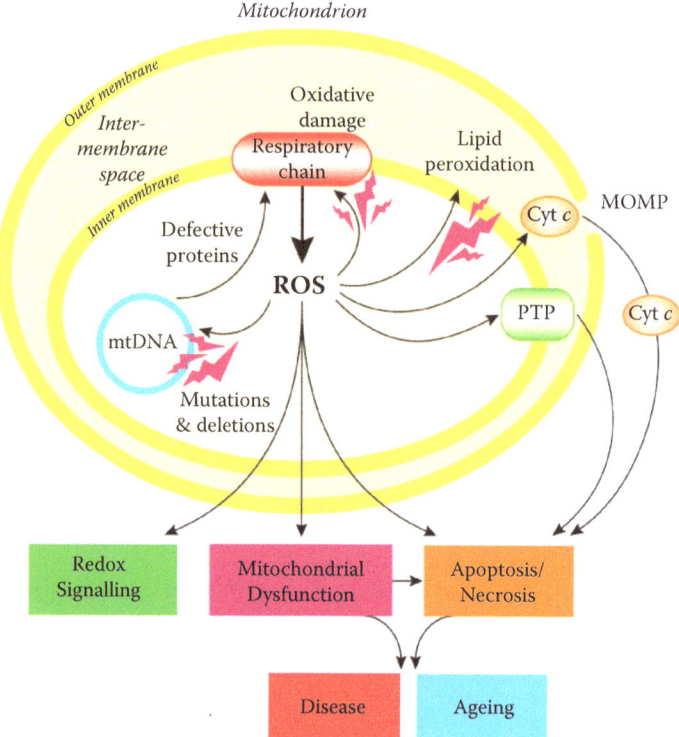

COLOR FIGURE 12.1 Overview of major cellular targets of oxidative damage. This diagram shows that electron leaks from the respiratory chain of mitochondria form ROS, which damage membrane lipids, generate defective proteins, and act as signaling molecules for opening permeability transition pore (PTP) and releasing membrane-bound cytochrome C (cytC). (From Murphy, *Biochem. J.* 417:1–13, 2009. Copyright © 2009, The Biochemical Society. Reproduced with permission.)

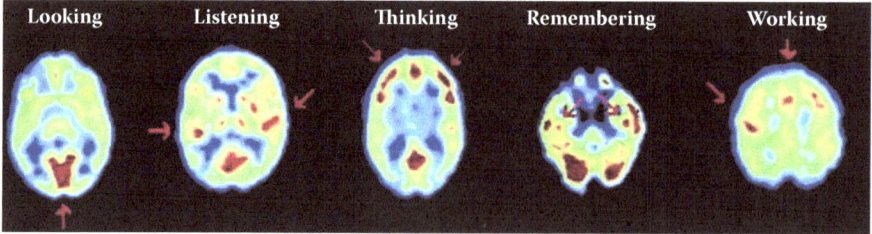

COLOR FIGURE 15.1 Positron emission tomography (PET) shows that metabolic activity and respiration in different regions of the brain are activated by different mental tasks. Since glucose consumed as a major energy source must be tightly coupled to lactate utilization by mitochondria, these data suggest that oxygen consumption and levels are also variable. Although a robust circulatory system feeds large amounts of oxygen to the brain, vast numbers of neurons and astrocytes use much of the oxygen, with the overall effect being a differential lowering of oxygen levels throughout the brain. (From Phelps, *Proc. Natl. Acad. Sci. USA* 97:9226–33, 2000. Copyright © 2000, National Academy of Sciences, U.S.A.)

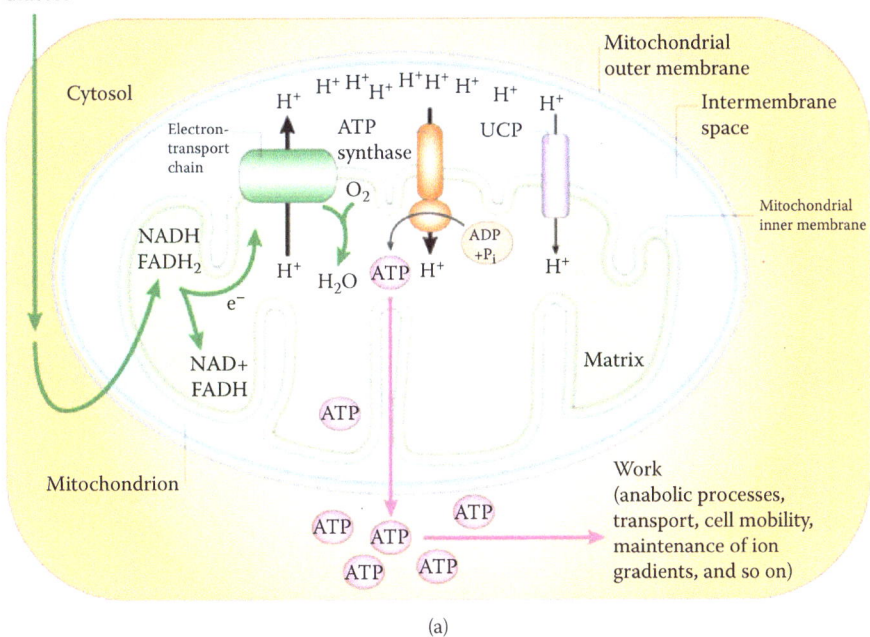

Fatty acids
Glucose

Cytosol

Mitochondrial
outer membrane

Intermembrane
space

Electron-
transport
chain

ATP
synthase

UCP

H^+ H^+ H^+ $H^+$$H^+$ H^+ H^+ H^+

H^+

Mitochondrial
inner membrane

O_2

ADP
$+P_i$

NADH
FADH$_2$

H^+ H_2O ATP H^+

H^+

e^-

NAD+
FADH

Matrix

ATP

Mitochondrion

ATP

ATP
ATP

ATP

Work
(anabolic processes,
transport, cell mobility,
maintenance of ion
gradients, and so on)

ATP ATP

(a)

COLOR FIGURE 16.1 UCPs 2–5 dissipate energy held in proton gradients and help protect the polyunsaturated fatty acid (PUFA)-enriched inner mitochondrial membrane against an oxidative chain reaction. (a) Diagram integrates mitochondrial bioenergetics and shows that UCPs act to uncouple the proton electrochemical gradients of mitochondria. See text for details. (From Krauss et al., *Nat. Rev. Mol. Cell. Biol.* 6:248–61, 2005. Copyright © 2005. Reprinted by permission from Macmillan Publishers Ltd.)

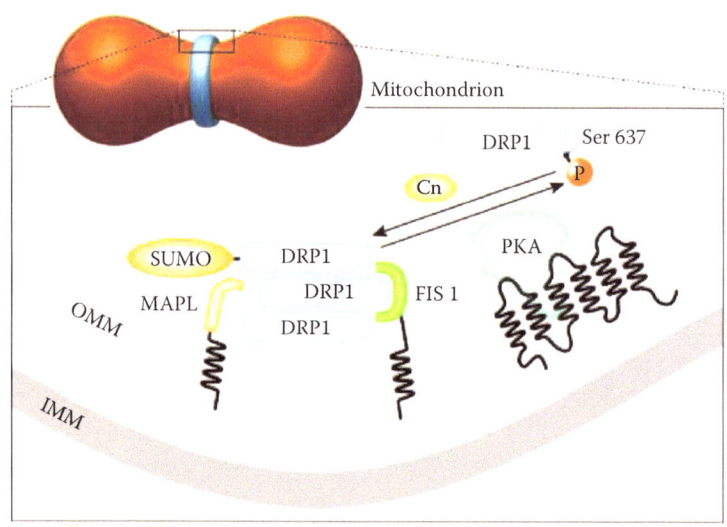

COLOR FIGURE 17.1 Diagram showing mechanism of formation of a fission "collar" generated by dynamic trafficking of dynamin-like protein DRP1 to the site of fission. A putative network regulating DRP1 accumulation and assembly is shown, in which DRP1 translocation is controlled by calcineurin-mediated dephosphorylation of Ser 637. Mitochondrial PKA (phosphokinase) then rephosphorylates the same site, pushing DRP1 away from the organelle. MAPL-mediated SUMOylation stabilizes DRP1 on mitochondria and might prevent its retranslocation to the cytoplasm. FIS1, fission 1; IMM, inner mitochondrial membrane; MAPL, mitochondrial-anchored protein ligase; OMM, outer mitochondrial membrane; PKA, protein kinase A; SUMO, small ubiquitin-like modifier. [Reprinted by permission from Macmillan Publishers Ltd: *EMBO Rep.* 10:694-6. Scorrano, L. and D. Liu. The SUMO arena goes mitochondrial with MAPL, © 2009.]

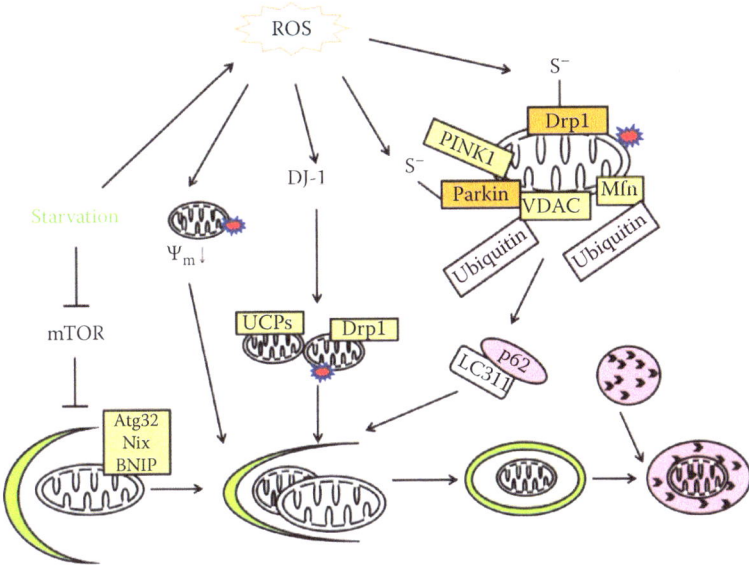

COLOR FIGURE 18.2 Oxidative stress can activate mitophagy. An elaborate global signaling network allows mammalian cells to tag and eliminate defective or unwanted mitochondria by the process of mitophagy. General signals include oxidative stress and starvation; specific signals include signaling proteins or modification of mitochondrial proteins. In yeast, Atg32 is specific for mitophagy and acts by targeting mitochondria to the autophagosome. In mammalian cells, Nix is involved in mitochondrial clearance during maturation of erythrocytes. During energy stress, when ATP levels drop, AMPK is activated and phosphorylates ULK1 and ULK2 (both Atg1 homologs) in turn activate mitophagy and general autophagy. Parkinson's disease genes encoding a-synuclein, parkin, PINK1, and DJ-1 are all involved in mitophagy. A decrease in mitochondrial membrane potential (ψ_m) can be induced by ROS and by targeting a-synuclein to the mitochondria. A depression in mitochondrial membrane potential serves as a signal for mitophagy. In addition to a decrease in membrane potential, mitochondrial fission is another signal for mitophagy. PINK1 facilitates parkin targeting to the mitochondria and ubiquitinates the mitochondrial outer membrane VDAC. Ubiquitinated VDAC can be recognized by p62 to initiate mitophagy. DJ-1 senses oxidative stress and serves as a parallel pathway to maintain mitochondrial membrane potential and preserve mitochondria from fragmentation. Many of the regulators of mitophagy can be modulated by ROS. For example, a-synuclein is nitrated and, as a consequence, increases aggregation propensity. Parkin can be sulfonated and S-nitrosated. Drp-1 S-nitrosation is also involved in regulation of mitochondrial fission and associated induction of mitophagy. (From Lee et al., *Biochem. J.* 441:523–40, 2012. Copyright © 2012, The Biochemical Society. Reproduced with permission.)

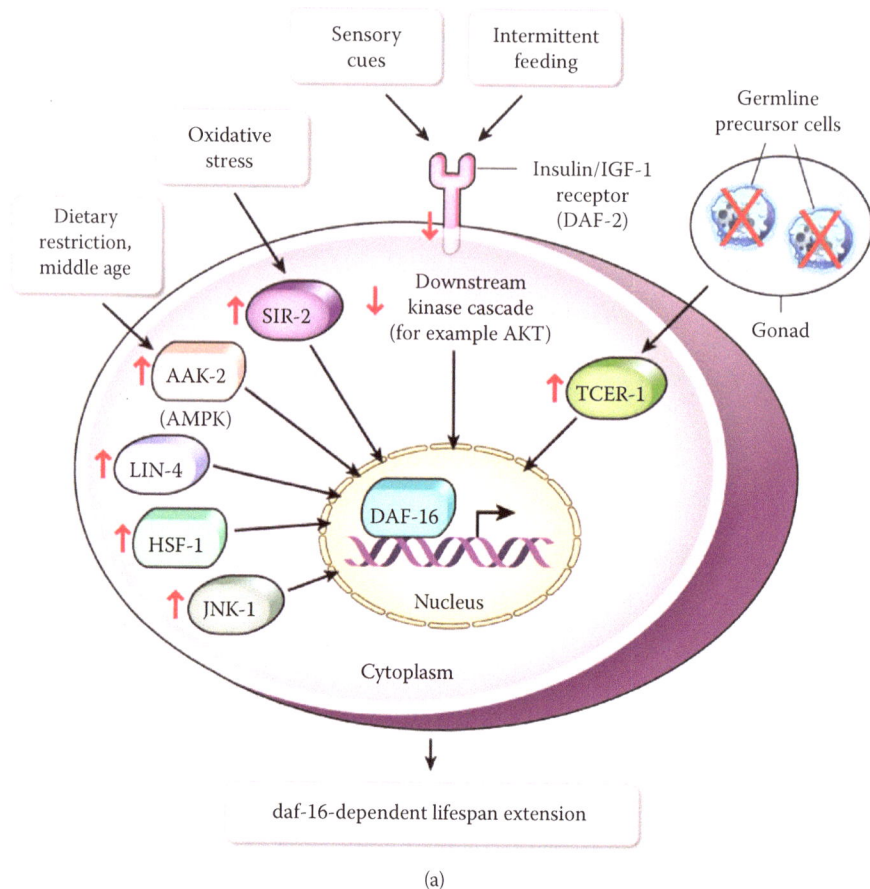

(a)

FIGURE 19.1 DAF-16 (FOXO-like transcription factor of *C. elegans*) regulates stress responses in *C. elegans*. (a) Overexpressing DAF-16, SIR-2 (sirtuin), HSF-1 (heat shock transcription elongation factor), LIN-4 (developmental-tuning micro RNA), AAK-2 (a subunit of AMP kinase), JNK-1 (JUN kinase), or the transcription elongation factor TCER-1 extends life span. Inhibiting the DAF-2 insulin/IGF1 receptor or components of its downstream kinase cascade also extends life span. In each case the extension of life span is DAF-16 dependent. (From Kenyon, *Nature* 464:504–12, 2010. Copyright © 2010. Reprinted by permission from Macmillan Publishers Ltd.) (b) Multiple FOXO transcription factors govern stress tolerance in mammals. (From Greer and Brunet, *Oncogene* 24:7410–25, 2005. Copyright © 2005. Reprinted by permission from Macmillan Publishers Ltd.)

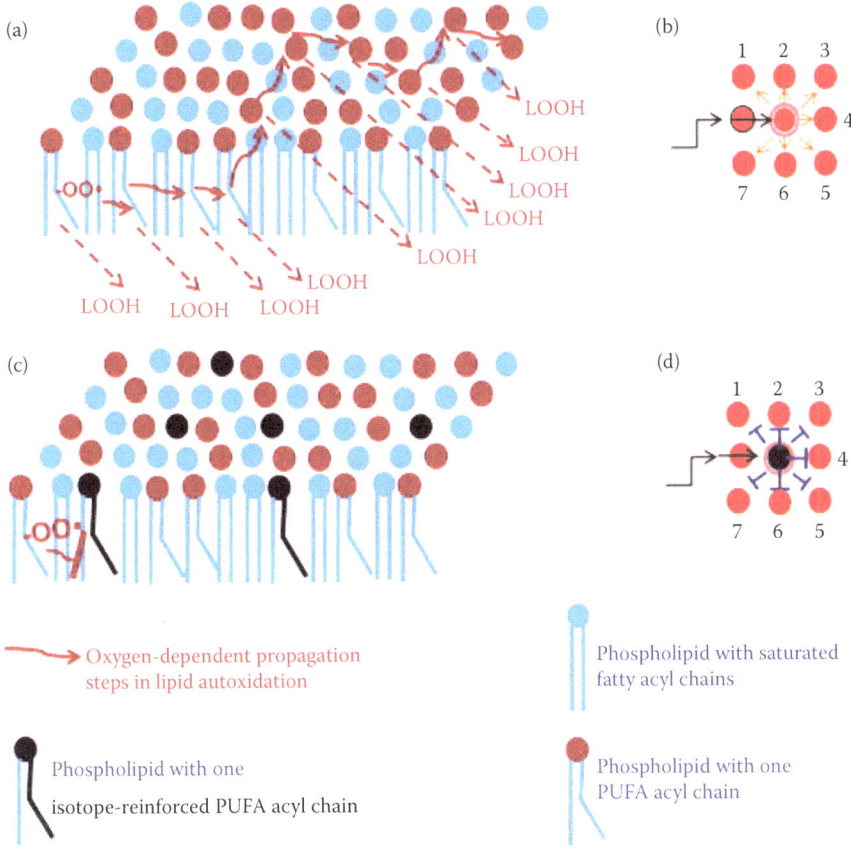

COLOR FIGURE 20.2 Model to explain how deuterium-reinforced PUFAs limit the chain reaction of membrane lipid peroxidation. (a) A theoretical chain reaction is depicted in a membrane where a single initiation event producing a lipid peroxyl radical (denoted by –OO•) starts a chain reaction of lipid autoxidation that in the presence of O_2 may continue indefinitely (red arrows) and produce many molecules of lipid peroxides. Susceptible phospholipid molecules containing a PUFA acyl chain are designated by a kinked blue line and a red dot. (b) Propagation of PUFA autoxidation can progress by interaction with any neighboring PUFAs. (c) The presence of 20 percent isotope-reinforced PUFA (denoted by a black kinked line and a black dot) inhibits (or slows) chain propagation. (d) Propagation is inhibited for PUFAs neighboring the D-PUFA. (From Hill et al., *Free Radic. Biol. Med.* 53:893–906, 2012. Copyright © 2012. Reprinted with permission from Elsevier.)

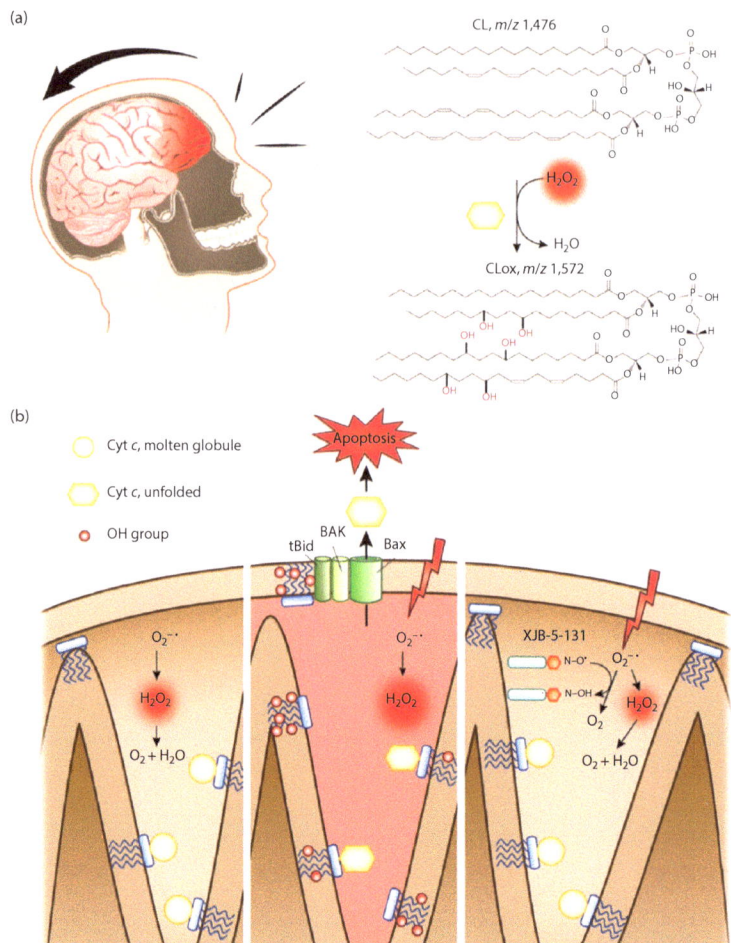

COLOR FIGURE 20.3 Cardiolipin (CL) as trigger molecule in the molecular pathology of traumatic brain injury (TBI) and discovery of mitochondria-targeted antioxidants protecting mitochondria. (a) Inflammation following TBI causes oxidation of highly unsaturated molecular species of cardiolipin, resulting in dysfunctional derivatives designated CLox. (b) Left panel shows reactive oxygen species such as H_2O_2 being detoxified and maintained at levels below a critical threshold for initiating a chain reaction. In the middle panel TBI overwhelms defenses against reactive oxygen species, unleashing an oxidative chain reaction propagated by oxidatively damaged CL. Oxidative membrane damage is so severe that membrane-bound cytochrome c is damaged and converted to its peroxidase derivative, further damaging the mitochondrial membrane. Cytochrome c escapes to the cytoplasm, triggering apoptosis. Right panel shows the antioxidant XJB-5-131 being targeted to the inner mitochondrial membrane. This membrane-targeted antioxidant intercepts sufficient numbers of superoxide radical ($O_2^{-\bullet}$), allowing superoxide dismutase to form H_2O_2, which is degraded by catalase yielding O_2 and H_2O. Thus, with the help of the membrane-targeted antioxidant, an oxidative chain reaction leading to cellular death is prevented. Note that the mechanism shown in the right panel can be generalized to cover oxidative stresses generated by aging, neurodegeneration, cancer, chronic inflammation, and mitochondrial diseases as well as TBI. (From Chan and Di Paolo, *Nat. Neurosci.*15:1325–27, 2012. Copyright © 2012. Reprinted by permission from Macmillan Publishers Ltd.)

Section III

Revised Mitochondrial Membrane Hypothesis of Aging

Recent data support the long-accepted view that aging is significantly a genetic event driven by accumulation over time of energy-robbing mutations in mitochondrial DNA. MtDNA is a circular molecule whose face can be envisioned as a sort of genetic clock. Mutations appear to occur randomly around the clock, and an important point is that multiple mutations in mtDNA likely govern aging. As mitochondria age and their capacity to produce energy declines, they continuously produce toxic reactive oxygen species (ROS). ROS can attack and further weaken mitochondria, often targeting their polyunsaturated membranes. Human cells have evolved many robust mechanisms to minimize ROS damage to membranes, but these defenses are energy-intensive (Section IV). We suggest that protecting mitochondrial membranes is bioenergetically far more costly than currently appreciated. Mitochondria, which accumulate mutations with aging, eventually produce less and less energy. The net effect is that an old mitochondrion diverts a greater proportion of its total energy output to protect its membranes than a young mitochondrion. Even if the energy cost to protect membranes remains constant with aging, the energy cost to protect membranes remains high and gobbles up an increasing amount of the cell's energy budget, causing energy stress. Thus, the amount of energy available for maintaining the health of the cell is hypothesized to decline toward a critical threshold, eventually triggering apoptosis.

8 Mitochondrial Diseases and Aging Have Much in Common

About 3000 genes are necessary for growth, maintenance, and recycling of mitochondria. Thirty-seven of these genes are encoded by mitochondrial DNA (Figure 8.1), with the nucleus harboring the remainder. Mutations in both mtDNA and nuclear genes cause mitochondrial diseases. Mitochondrial diseases most often affect children, and many are lethal. However, adult-onset or age-dependent mitochondrial diseases are becoming more common. Dysfunctional mitochondria appear to cause the most damage to cells of the brain, heart, liver, skeletal muscles, kidney, and endocrine and respiratory systems. Some symptoms of mitochondrial diseases include muscle weakness and pain, loss of motor control, gastrointestinal disorders, swallowing difficulties, poor growth, cardiac disease, liver disease, diabetes, respiratory complications, seizures, visual and hearing problems, mild to fatal lactic acidosis, developmental delays, and susceptibility to infection. This wide range of symptoms is not surprising given that mitochondria produce more than 90 percent of the energy needed by the body to sustain life.

Mitochondrial diseases can be classified by symptoms, timing of onset, inheritance, class of mutations, mitochondrial versus nuclear genes, organ specificity, and so forth. The focus here is on selected case histories of mitochondrial diseases in which key mitochondrial enzymes, damaged by mutations in either mtDNA or nuclear genes, cause energy stress followed by disease symptoms in an age-dependent manner.

8.1 LEBER'S HEREDITARY OPTIC NEUROPATHY (LHON)

LHON is a hereditary disease transmitted from the mother to all offspring as a genetic defect or mutation in mtDNA (Yu-Wai-Man et al., 2011). In 90 percent of cases only the egg contributes mutated mitochondria to the embryo. LHON is due to one of three mtDNA point mutations (i.e., at positions 11778, 3460, and 14484). Note that mutations causing LHON occur in ND1, ND4, ND4L, and ND6 subunit genes encoding complex 1 of the electron transport chain of human mitochondria (Figure 8.1). ND is shorthand for NADH dehydrogenase, which is a large electron transport complex embedded in the mitochondrial inner membrane and sitting at the beginning of the electron transport chain. This powerful proton pump is an extremely sophisticated membrane-bound enzyme assembled from a remarkable number of subunits, seven of which are encoded by genes of mtDNA.

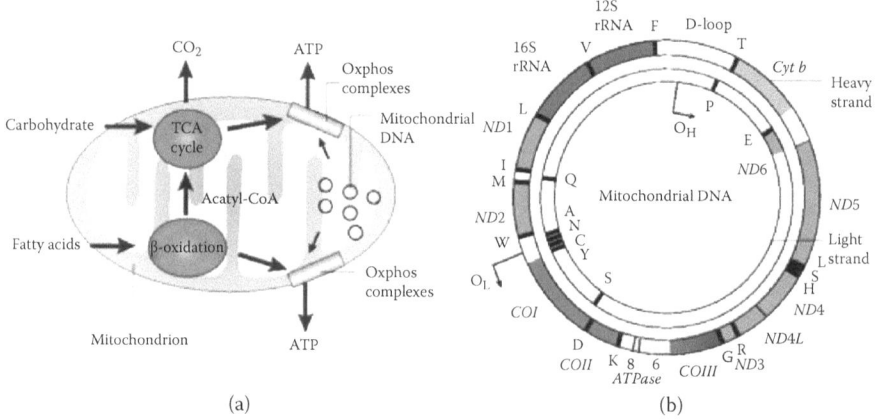

FIGURE 8.1   See color insert. Human mitochondrial DNA (mtDNA) is a small, circular double helix encoding several essential energy-producing genes. The genes that encode the subunits of complex 1 (ND1–6 and ND4L) are shown in blue; the terminal complex, cytochrome c oxidase (COI-COIII), is shown in red; cytochrome b of complex 3 is shown in green; and the subunits of ATP synthase (ATPase 6 and 8) are shown in yellow. RNA genes are also listed (purple and black slashes). (From Taylor and Turnbull, *Nat. Rev. Genet.* 6:389–402, 2005. Copyright © 2005. Reprinted by permission from Macmillan Publishers Ltd.)

Complex 1 handles high-energy electrons derived from foodstuffs and shuttled to complex 1 by the universal electron carrier NADH; electrons donated by NADH are used to energize the efflux of protons out of the matrix of mitochondria as catalyzed by complex 1 (Hirst, 2010; Pryde and Hirst, 2011). Mutations that completely block the proton pumping action of NADH dehydrogenase are lethal in contrast to point mutations such as occur in LHON, which are slow but do not completely block activity of this crucial enzyme. These data suggest that LHON is an energy deficiency or energy stress disease. It is clear that lowering ATP output caused by mutations in complex 1 or other mechanisms can give rise to a diverse set of symptoms affecting selected, but not all, cells and organs of the human body. There are numerous unexplained features of LHON, such as its effects on predominantly young adult males, onset of symptoms ranging from age eight to sixty years, severity of symptoms, synergism with environmental factors, including light, and the appearance of the rare LHON-plus state of the disease. This higher form of the disease is characterized by lack of muscle control, tremors, and cardiac arrhythmia. One of the most surprising features of LHON is that about 50 percent of males carrying the mutation and 85 percent of females with the mutation never experience visual loss or related medical problems. When visual problems do appear, they are limited to the retinal ganglion cell layer where degeneration occurs via apoptosis.

In a revealing case history concerning impact of the environment on LHON, a linkage has been established between visible light and degeneration of retinal ganglion cells. This relationship might be explained in part by a membrane photooxidation mechanism such as occurs in rhodopsin disks (see Chapter 12). In a comprehensive overview of the molecular biology of LHON, Yu-Wai-Man and colleagues (2011) point out that mitochondrial membranes in nonmyelinated retinal ganglion axons are directly exposed to light. Mitochondrial membranes of these specialized neurons seem to be targets of photooxidation. Details on the role of light and other factors in predisposing retinal ganglion cells of patients with LHON to premature apoptosis are discussed by Yu-Wai-Man et al. (2011). Is LHON caused by oxidative stress or energy stress?

Recently Lin and coworkers (2012) developed a mouse model of LHON by introducing the human optic atrophy mtDNA ND6P25L mutation into the mouse. Mice with this mutation exhibited many of the expected pathologies of LHON, as follows:

- Reduction in retinal function
- Age-related decline in central smaller-caliber optic nerve fibers
- Sparing of larger peripheral fibers
- Neuronal accumulation of abnormal mitochondria
- Axonal swelling
- Demyelination

Mitochondria displayed partial defects in complex 1 and rates of respiration. Levels of reactive oxygen species (ROS) increased. Synapse analysis revealed decreased complex 1 activity and increased ROS but no drop in ATP production. Lin and colleagues (2012) conclude that LHON pathology might result from oxidative stress in contrast to energy stress.

The Hirst model (Hirst, 2010; Birrell et al., 2009) for production of ROS at the high-energy (flavin mononucleotide (FMN)) site of complex 1 (see Chapter 11) might account for increased ROS production in the mouse model of LHON. According to this concept, a restriction in normal electron flow through complex 1 is expected to increase ROS production because of the increase in the ratio of NADH:NAD. This ratio governs the electron saturation level of the high-energy site of complex 1 with higher NADH levels, causing an increased flux of electrons to form superoxide.

We offer a word of caution concerning the interpretation of the data on ATP production that suggests that oxidative, not energy, stress is the root cause of LHON. We consider the comparison of mitochondria of mouse versus man similar to comparing apples to oranges because of their dramatic differences in fatty acid composition and membrane lipid dynamics. That is, mouse mitochondria are highly enriched with docosahexaenoic acid (DHA) and other highly unsaturated fatty acids (HUFAs) compared to those of man. We suggest that Hirst's model helps explain the data of Lin and colleagues (2012). Perhaps the normally high DHA levels in mouse mitochondria maintain high levels of ATP. Thus, we suggest that the data of Lin and coworkers (2012) may not clearly distinguish between oxidative stress and energy stress as the cause of LHON in humans. (Also see Hagopian et al. (2010) and Chapter 11 for another case history in which the Hirst model helps explain data on the relationship between respiration and oxidative stress in mice.)

8.2   BARTH'S SYNDROME

Cardiolipin (CL), the signature lipid in mitochondrial membranes, plays multiple biochemical functions and is implicated in numerous human pathologies (Chicco and Sparagna, 2007) including aging (see Chapters 9 and 20). The essential roles of CL in energy transduction in mitochondria of mammals were established using CL knockout mutants of mice (Osman et al., 2010; Zhang et al., 2011) and biochemical-genetic analysis of Barth's syndrome (Schlame and Ren, 2006).

Barth's syndrome, also called lethal infantile cardiomyopathy, is considered a membrane disease because cardiolipin homeostasis in the mitochondrial membranes is mutationally disturbed. The inherited x-linked mutation, striking males only, does not completely block the synthesis of CL. Rather, the mutation changes its structure, with the result of generating abnormal molecular species. The major molecular species of CL in cardiac cells is $(18:2)_4$-CL, in which linoleic acid (18:2) occupies all four acyl positions. In Barth's syndrome the trend is that the 18:2 chains are replaced with more saturated fatty acids, including 18:1. The molecular species $(18:1)_4$-CL is considered an immature species of CL, but if this more saturated structure persists, then disease symptoms follow.

Barth's syndrome is caused by mutations in the tafazzin gene. Tafazzin is essential in remodeling of newly synthesized CL, functioning as a monolysocardiolipin transacylase (see Chapter 6). Remodeling is initiated by a CL-specific deacylase that removes one acyl chain and generates a CL structure with only three chains. Mutations in tafazzin cause the three-acyl derivative to accumulate, as the remodeling pathway is initiated but cannot be completed. We suggest that CL missing one of its acyl chains might cause membrane dysfunction and behave as an energy uncoupler in mitochondria.

Numerous mutations in TAZ1 are associated with Barth's syndrome, including frame shift mutations that cause tafazzin truncation and mutations affecting messenger RNA splicing. In a recent study, twenty-one of a total of twenty-eight missense mutations in tafazzin that cause Barth's syndrome were modeled in yeast (Claypool et al., 2011). These data show that loss of function of eighteen of twenty-one mutants tested is due to the inherent instability of the mutant tafazzin complexes. Misfolding of tafazzin caused by missense mutations is believed to be monitored by the protein quality control apparatus of the inner mitochondrial membrane, resulting in proteolysis of these dysfunctional proteins. Thus, mutational analysis of tafazzin shows at least three critical biochemical properties are responsible for loss of function of tafazzin—dysfunctional stability, aberrant targeting, and improper assembly.

Studies of Barth's syndrome establish the essential role of cardiolipin in energy transduction in human mitochondria. CL is primarily located in the inner membrane but is distributed in both leaflets. Roughly 10 to 15 percent of the total surface of the inner membrane is estimated to be composed of CL. The universality, localization, amounts, and structure of CL suggest multiple beneficial roles, including formation of supramolecular complexes enhancing energy efficiency as well as appropriate membrane motion and permeability. These benefits are balanced against risks associated with membrane integrity and oxidative stability. We suggest that a tripartite membrane fatty acid blending code (Chapter 3) is applicable to understanding the biochemical roles of CL. For example, $(18:2)_4$-CL is expected to help maximize energy production contributing antirafting properties and maintaining membrane homeostasis. Another fundamental role of mitochondrial membranes involves lipid architecture tight enough to block spontaneous passage of protons. We propose that mitochondrial fatty acid structure has evolved to maximize respiratory energy production and energy conservation balanced against risks of dysfunctional membranes caused by peroxidation. As discussed in Chapter 4, CL in mitochondria of nematodes is specifically targeted and degraded as a function of aging. Thus, CL seems to be a reporter of membrane peroxidation as well as the aging process, presumably by an oxidative mechanism.

As discussed above, numerous biochemical roles have been suggested for CL, and a new role is now receiving a great deal of attention. The field of CL-induced membrane curvature as a mechanism to attract specific proteins to curved membrane regions of cells or organelles has recently been reviewed by Huang and Ramamurthi (2010). This lucid and thought-provoking review includes a comprehensive reference list that is a "must read" in this field. In reading this review, we were struck by the powerful role played by interdisciplinary teams of scientists tackling this difficult question of membrane curvature function. The data are convincing that a new dimension in the field of membrane lipid structure-function has opened up, with major implications for understanding the contributions of cardiolipin in health and disease. There is also a certain personal satisfaction for us in that bacteria and microorganisms as research tools continue to play a part in deciphering a new biochemical role proposed for cardiolipin in membrane curvature (Renner and Weibel, 2011).

8.3   LATENT MITOCHONDRIAL DISEASES CAUSED BY MUTATIONS IN THE MTDNA-REPLICATING MACHINE (POLG) MAY NOT DISPLAY SYMPTOMS FOR UP TO SIXTY YEARS

Numerous mitochondrial diseases are caused by mutations in the POLG gene, which encodes the catalytic subunit of the only known mtDNA replicase in mitochondria (Wong et al., 2008). Mitochondria often divide every ten days or so (Chapter 17), keeping POLG busy replicating new copies of mtDNA. POLG not only acts as a DNA replicating machine, but also carries out proofreading of the newly synthesized mtDNA strand. The POLG gene is encoded by nuclear DNA, and the protein product is subsequently targeted to mitochondria. Mutations have occurred spontaneously in the general population, causing structural changes affecting the dynamics of mtDNA replication, including the frequency at which errors of replication (mutations) occur. Mutations that increase the levels of errors of replication convert POLG into its mutator form discussed in Chapter 9. Mutations in POLG appear to affect all aspects of the structure-function of this enzyme, including translocation rates from the nucleus to mitochondria and stability. The net effect of mutational damage of POLG in humans is an almost bewildering spectrum of neurodegenerative diseases, as follows:

- Childhood myocerebrohepatopathy spectrum (MCHS) disorders
- Alpers' syndrome

- Ataxia neuropathy spectrum (ANS) disorders
- Myoclonic epilepsy myopathy sensory ataxia (MEMSA)
- Autosomal recessive progressive external ophthalmoplegia (arPEO)
- Autosomal dominant progressive external ophthalmoplegia (adPEO)

There are numerous other ways of classifying POLG-mediated mitochondrial diseases, including their frequency and severity. For example, Alpers' syndrome appears to be the most common autosomal recessive disease caused by mutations in the POLG gene. This fatal disease is characterized by intractable seizures, hepatic failure, and global neurological deterioration. Note that two specific organs, mainly the liver and nervous system, are destroyed during the progression of Alpers' syndrome. Biochemists consider mitochondria in these two organs to be among the hardest working in the body based on their rapid turnover rates (e.g., about ten days for liver). This pattern of specific organ failure has been observed with other mitochondrial diseases and includes the heart in addition to the brain and liver. POLG-mediated mitochondrial diseases can also be classified based on the number of POLG mutations involved. Both parents can contribute a separate POLG mutation to their offspring. The finding that multiple mutations in POLG can cause mitochondrial diseases has implications for aging, where multiple mutations in genes encoded in mtDNA are proposed to cause energy stress and drive the aging process (see Chapter 9).

Many POLG mutations cause early-onset fatal diseases, but late-onset diseases are being increasingly recognized (Table 8.1). Table 8.1 summarizes the age of onset of diagnosed POLG-mediated disease in thirty patients. Note that MCHS had the earliest age of onset (one year). Twenty of the thirty patients in this study had Alpers'

TABLE 8.1

Mutator Genes (POLG Mutations) in Humans Can Generate Age-Dependent Mitochondrial Diseases

| Diagnosis | N = | Median | Age of Onset (years) | | Mean | SD | Min. | Max. |
			10th Percentile	90th Percentile				
MCHS	3	1.0	—	—	1.4	3.5	0.2	3
Alpers'	20	2.0	0.9	9.5	3.6	5.2	0.5	23
ANS	3	17	—	—	16	1.2	15	17
arPEO+	4	40	—	—	43	13	32	60
adPEO+	3	46	—	—	46	21	25	66
Unassigned 3								
Total	36							

Source: Wong et al., *Hum. Mutat.* 29:E150–72, 2008. Copyright © 2008. Reprinted with permission from John Wiley & Sons.

Note: —, insufficient data for calculation.

syndrome, with a median age of onset of two years. The median age of onset of ANS was seventeen years, compared to forty years for arPEO and forty-six for adPEO. The maximum age of onset for arPEO was sixty years, compared to sixty-six years for adPEO.

Interestingly, a conservative amino acid substitution involving a shift from isoleucine to leucine within the polymerase domain of POLG was identified in the sixty-six-year-old patient (Wong et al., 2008). The critical catalytic domain of POLG where this amino acid shift was noted is conserved across all life-forms examined so far, with isoleucine seen in all cases with the exception of *Drosophila*. POLG of *Drosophila* has leucine in place of isoleucine, matching the isoleucine → leucine substitution seen in the late-onset patient. This shift involving two very similar amino acids is called a conservative change because both amino acids are believed to play nearly identical structural and catalytic roles in enzyme catalysis. In other words, a shift from isoleucine to leucine is considered here to be a subtle mutational change because it is difficult using any biochemical assay available today to detect any change in function of the altered enzyme. However, no biochemical assay that we are aware of has the sensitivity associated with an incubation time of sixty-six years (as seen for the onset of symptoms for the patient carrying the POLG isoleucine → leucine mutation).

Thus, aging itself can be considered a bioassay for monitoring the effects of subtle mutational changes on a timescale previously unimaginable by biochemists. Studies of mutations causing aging and age-related diseases are motivating biochemical geneticists to reevaluate the nature of neutral human mutations to include a timescale of 0 to 100 years. We propose that subtle mutations similar to the isoleucine → leucine shift will turn out to play important roles in understanding age-dependent diseases and aging itself.

8.4   FRIEDREICH'S ATAXIA

Friedreich's ataxia (FRDA) is a devastating orphan disease affecting about 3 in 100,000 individuals in Caucasian populations. FRDA is a mitochondrial disease. It is caused by intronic GAA repeat expansions that hinder expression of the frataxin (FXN) gene encoded in the nucleus. Down-regulation results in defective levels of the mitochondrial-targeted protein frataxin. There is no specific therapy for FRDA, but considerable effort is now being devoted to finding a cure. For example, interferon gamma has been found to up-regulate frataxin and correct functional defects in a mouse model of FRDA. Other possible therapies are being explored, including the use of deuterated polyunsaturated fatty acids (see Chapter 20) as protective agent (Grazia Cotticelli et al., 2013). As a result of this intense research activity, the molecular pathology of FRDA is among the best understood among mitochondrial diseases, and there is increasing optimism that a treatment can be found. See Grazia Cotticelli and colleagues (2013) for references concerning potential treatments for FRDA.

It is well known that FRDA is an age-dependent disease. This means that Friedreich's ataxia likely shares some fundamental mechanisms with aging (Vyas

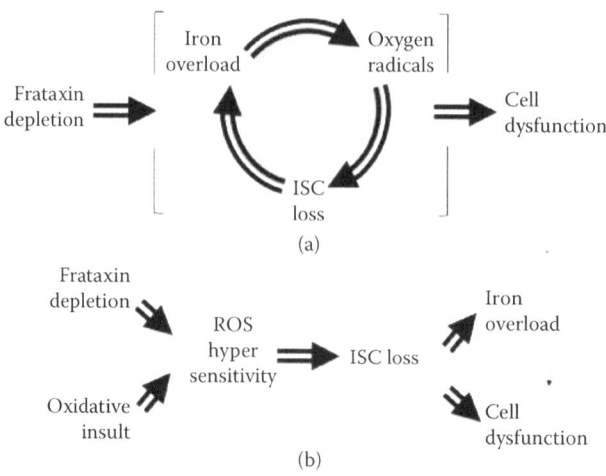

FIGURE 8.2 The vicious circle hypothesis of Friedreich's ataxia features mitochondria predisposed to hypersensitivity to ROS. (a) According to the vicious circle hypothesis, frataxin depletion results in impaired iron-sulfur cluster synthesis or stability with intramitochondrial accumulation of reactive iron. Reactive iron is envisioned to promote Fenton chemistry, producing superoxide and hydrogen peroxide, which in turn destroys more iron-sulfur clusters (ISCs). (b) In frataxin-depleted cells, deficient signaling of antioxidant defenses is proposed to sensitize the frataxin-free iron-sulfur clusters to destruction by ROS. (From Bayot et al., *BMC Med.* 9:112, 2011. Copyright © 2011, BioMed Central Ltd.)

et al., 2012; Koeppen, 2011; Koeppen et al., 2012; Xia et al., 2012; Lill et al., 2012). Recently, Bayot and colleagues (2011) revisited an earlier hypothesis, the vicious circle concept (Figure 8.2). These researchers point out that data from cell and animal models now indicate that iron accumulation, previously thought to be a marker of this disease, is an inconsistent and late event. They also suggest that frataxin deficiency does not always impair the activity of iron-sulfur cluster-containing proteins essential in the electron transport chain of mitochondria. As a general rule, frataxin deficiency appears to be associated with increased sensitivity to reactive oxygen species as opposed to elevated levels of oxygen radicals. Bayot and coworkers (2011) suggest that frataxin deficiency creates an abnormal oxidative status that triggers the pathogenic mechanism underlying Friedreich's ataxia (Figure 8.2). This is a provocative model leading to the idea that hypersensitivity to oxygen radicals might be a more reliable reporter of events in FRDA than the negative data showing the absence of a spike in ROS levels. For example, a scenario can be envisioned in which ROS levels rise simultaneously with up-regulation of ROS defenses, with the net effect being an increased probability of a chain reaction with perhaps a barely detectable rise in ROS levels. Time also enters this picture in that while ROS levels might rise only marginally or below detection levels, over time oxidative damage to mitochondria might occur. Once again, the chicken-egg story of which comes first—oxidative stress or energy stress—is raised.

8.5 SUMMARY

Selected mitochondrial diseases were chosen here to illustrate common threads among mitochondrial disease, energy stress, oxidative stress, and aging, as follows:

- Data from studies of LHON established the importance of mitochondrial diseases in humans.
- Studies of Barth's syndrome established the essential role of cardiolipin in energy transduction by human mitochondria, data that support multiple biochemical roles of CL in energy stress, oxidative stress, and aging.
- Analysis of human POLG mutations is revealing a gold mine of information on the nature of mutations causing age-dependent diseases with implications for aging itself (also see Chapter 9).
- Understanding of Friedreich's ataxia has reached the stage that there is optimism that this disease might be cured in the immediate future.

Finally, mitochondrial diseases continue to play pivotal roles in understanding and integrating the linkage between mitochondrial bioenergetics, oxidative stress, and aging.

REFERENCES

Bayot, A., R. Santos, J. M. Camadro, et al. 2011. Friedreich's ataxia: the vicious circle hypothesis revisited. *BMC Med.* 9:112.

Birrell, J. A., G. Yakovlev, and J. Hirst. 2009. Reactions of the flavin mononucleotide in complex I: a combined mechanism describes NADH oxidation coupled to the reduction of APAD+, ferricyanide, or molecular oxygen. *Biochemistry* 48:12005–13.

Chicco, A. J., and G. C. Sparagna. 2007. Role of cardiolipin alterations in mitochondrial dysfunction and disease. *Am. J. Physiol. Cell Physiol.* 292:33–44.

Claypool, S. M., K. Whited, S. Srijumnong, et al. 2011. Barth syndrome mutations that cause tafazzin complex lability. *J. Cell Biol.* 192:447–62.

Grazia Cotticelli, M., A. M. Crabbe, R. B. Wilson, et al. 2013. Insights into the role of oxidative stress in the pathology of Friedreich ataxia using peroxidation resistant polyunsaturated fatty acids. *Redox Biol.* 1:398–404.

Hagopian, K., K. L. Weber, D. T. Hwee, et al. 2010. Complex I-associated hydrogen peroxide production is decreased and electron transport chain enzyme activities are altered in n-3 enriched fat-1 mice. *PLoS One* 5(9):e12696.

Hirst, J. 2010. Towards the molecular mechanism of respiratory complex I. *Biochem. J.* 425:327–39.

Huang, K. C., and K. S. Ramamurthi. 2010. Macromolecules that prefer their membranes curvy. *Mol. Microbiol.* 76:822–32.

Koeppen, A. H. 2011. Friedreich's ataxia: pathology, pathogenesis, and molecular genetics. *J. Neurol. Sci.* 303:1–12.

Koeppen, A. H., R. L. Ramirez, D. Yu, et al. 2012. Friedreich's ataxia causes redistribution of iron, copper, and zinc in the dentate nucleus. *Cerebellum* 11:845–60.

Lill, R., B. Hoffmann, S. Molik, et al. 2012. The role of mitochondria in cellular iron-sulfur protein biogenesis and iron metabolism. *Biochim. Biophys. Acta* 1823:1491–508.

Lin, C. S., M. S. Sharpley, W. Fan, et al. 2012. Mouse mtDNA mutant model of Leber hereditary optic neuropathy. *Proc. Natl. Acad. Sci. USA* 109:20065–70. doi: 10.1073/pnas.1217113109.

Osman, C., M. Haag, F. T. Wieland, et al. 2010. A mitochondrial phosphatase required for cardiolipin biosynthesis: the PGP phosphatase Gep4. *EMBO J.* 29:1976–87.

Pryde, K. R., and J. Hirst. 2011. Superoxide is produced by the reduced flavin in mitochondrial complex I: a single, unified mechanism that applies during both forward and reverse electron transfer. *J. Biol. Chem.* 286:18056–65.

Renner, L. D., and D. B. Weibel. 2011. Cardiolipin microdomains localize to negatively curved regions of *Escherichia coli* membranes. *Proc. Natl. Acad. Sci. USA* 108:6264–69.

Schlame, M., and M. Ren. 2006. Barth syndrome, a human disorder of cardiolipin metabolism. *FEBS Lett.* 580:5450–55.

Taylor, R. W., and D. M. Turnbull. 2005. Mitochondrial DNA mutations in human disease. *Nat. Rev. Genet.* 6:389–402.

Vyas, P. M., W. J. Tomamichel, P. M. Pride, et al. 2012. A TAT-frataxin fusion protein increases lifespan and cardiac function in a conditional Friedreich's ataxia mouse model. *Hum. Mol. Genet.* 21:1230–47.

Wong, L. J., R. K. Naviaux, N. Brunetti-Pierri, et al. 2008. Molecular and clinical genetics of mitochondrial diseases due to POLG mutations. *Hum. Mutat.* 29:E150–72.

Xia, H., Y. Cao, X. Dai, et al. 2012. Novel frataxin isoforms may contribute to the pathological mechanism of friedreich ataxia. *PLoS One* 7(10):e47847.

Yu-Wai-Man, P., P. G. Griffiths, and P. F. Chinnery. 2011. Mitochondrial optic neuropathies: disease mechanisms and therapeutic strategies. *Prog. Retin. Eye Res.* 30:81–114.

Zhang, J., Z. Guan, A. N. Murphy, et al. 2011. Mitochondrial phosphatase PTPMT1 is essential for cardiolipin biosynthesis. *Cell. Metab.* 13:690–700.

9 Revised Mitochondrial Hypothesis of Aging Highlights Energy Deficiency Caused by Errors of Replication (Mutations) of mtDNA

Energy is the basis of all human activity, ranging from heavy lifting to intense thinking. An estimated total of 10 quadrillion mitochondria act as power sources in human cells. These bacteria-like organelles produce the primary energy currency of all cells, mainly proton electrochemical gradients yielding ATP. The process of converting food to useful energy by mitochondria is called respiration and consumes most of the oxygen carried by our circulating red blood cells. Mitochondria likely originated from bacteria or archaea, but long ago lost their free-living lifestyle, gaining in its place a permanent and bioenergetically favorable endosymbiotic relationship within animal and plant cells. These energy-transducing organelles, which still retain some properties of bacteria, are strictly dependent on their host cell for growth, repair, and recycling, but have retained a miniature chromosome of their own called mitochondrial DNA (mtDNA) (see Chapter 8).

MtDNA encodes several genes essential for energy production, and in a strange quirk of nature these genes appear to hold life and death powers over the human organism—acting as a master pacemaker for aging and brain span (Kukat et al., 2011; Amati-Bonneau et al., 2008; Seo et al., 2010; Nakada et al., 2009; Paradies et al., 2011). In this chapter we explore how replication errors of mtDNA are linked to energy stress and aging in mice and neurodegeneration in humans. A mitochondrial, fusion-based mechanism pushing back against energy stress is also discussed. Interestingly, a genetically engineered mutator gene was found to accelerate aging in mice and occurs naturally in humans (see Chapter 8), where it causes premature death of neurons. This suggests that energy stress can strike the brain selectively, in essence fast-forwarding aging of the brain.

9.1 MTDNA ENCODES SEVEN SUBUNITS OF COMPLEX 1

Each mitochondrion is estimated to contain two to ten mtDNA copies. In humans 1000 to 10,000 separate copies of mtDNA are present per cell. Human mitochondrial

DNA (Chapter 8) is composed of 16,569 base pairs, or twice this many nucleotides, forming the famous double-helix structure. MtDNA is shaped as a closed circular molecule. The two intertwined strands of mtDNA that make up the double helix are referred to as the heavy strand and the light strand. They each have distinct nucleotide compositions. The heavy strand encodes twenty-eight genes compared to nine genes for the light strand of the double helix, for a total of thirty-seven. Thirteen genes code for proteins (polypeptides), all of which play primary roles in energy production. Seven of these genes encode different subunits of complex 1, the initial proton pump in the electron transport chain receiving its electrons from NADH (Figure 9.1). We will see later that genes for complex 1, also called NADH dehydrogenase, represent especially large targets for mutations, mutations proposed to act as pacemakers of aging. Mutationally modifying this powerful proton pump can affect energy homeostasis and even cause death (Chapter 8). Mutations in many other genes also modulate energy production and energy homeostasis, and the numbers keep climbing. For example, two subunits of ATP synthase, the rotary enzyme that harnesses proton electrochemical fuel for production of ATP, are encoded by mtDNA, and energy production can be modulated by mutations in these genes. Genes encoding subunits of other essential components are also present, along with twenty-four genes encoding for RNA products needed for protein synthesis essential for mitochondrial division and for replacing proteins, which tend to wear out rapidly in mitochondria.

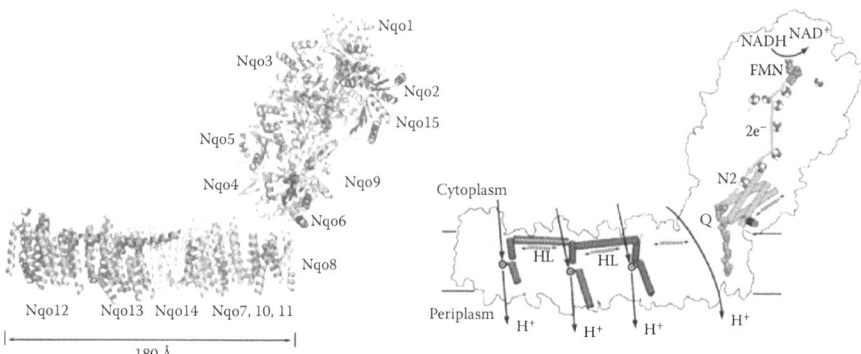

FIGURE 9.1   See color insert. Structure of chair-shaped complex 1 showing numerous subunits (left) and topology of electron flow (right). Note that electrons flow from NADH as donor through iron-sulfur centers located in the long arm of the enzyme extending into the mitochondrial matrix. The proton-pumping portion of the enzyme is embedded in the membrane. High-energy electrons flowing from NADH to ubiquinone lose energy while energizing the efflux of protons. Complex 1, composed of forty-five different subunits, is by far the most complicated member of the electron transport chain and is a major target of mutations occurring in mtDNA. Complex 1 is the largest enzyme whose structure has been solved, and this achievement opens a new window for understanding membranes' contribution to longevity. (From Efremov et al., *Nature* 465:441–45, 2010. Copyright © 2010. Reprinted by permission from Macmillan Publishers Ltd.)

New copies of mtDNA are produced during mitochondrial division or fission by a replicating machine called the DNA polymerase gamma enzyme complex, introduced in Chapter 8. This enzyme is composed of a larger catalytic DNA polymerizing enzyme encoded by the POLG gene and a smaller subunit coded for by the POLG2 gene. Note that nuclear genes encode these DNA-replicating enzymes, which are sent or targeted from the nucleus into mitochondria. Placing DNA polymerase genes in the nucleus rather than being encoded by mtDNA might have the effect of protecting these essential genes from high levels of mutations seen in mitochondria (Longley et al., 2005). The revised mitochondrial hypothesis of aging discussed next features energy stress as the direct cause of aging and is based on studies with a now famous line of mice—mutator mice.

9.2   MUTATED POLG IN MICE ACCELERATES AGING

A great deal of correlative evidence linking mitochondrial dysfunction and aging has emerged over the last several decades (reviewed by Trifunovic and Larsson, 2008). The creation of mtDNA mutator mice provides the clearest picture yet that fast-forwarding mtDNA mutations causes premature aging (Trifunovic et al., 2004; Dogan and Trifunovic, 2011). Thus, the first direct evidence is available showing that mutational loss of mitochondrial energy production is a major causal factor in mammalian aging. Mutator mice were genetically engineered with a defective catalytic subunit of mitochondrial DNA polymerase (POLGA), which speeds up random accumulation of mtDNA mutations. Mutations in mtDNA of mutator mice accumulate more rapidly than wild-type controls. Studies of mutator mice show that increased levels of mutations in mtDNA and subsequent energy stress can directly cause or fast-forward numerous age-related symptoms in mice (Trifunovic et al., 2004; Trifunovic and Larsson, 2008; Tyynismaa and Suomalainen, 2009; Edgar and Trifunovic, 2009), as follows:

- Weight loss
- Progressive hearing loss
- Heart disease
- Osteoporosis
- Reduced subcutaneous fat
- Alopecia
- Kyphosis
- Anemia
- Sarcopenia
- Reduced fertility
- Decreased spontaneous activity

The mtDNA mutator mice appear completely normal at birth and early adolescence, but subsequently display many features of premature aging now considered to be caused by energy stress. For example, hearing loss occurs at roughly double the rate in mutator mice as in controls (Niu et al., 2007; Someya et al., 2008). At the cellular level mtDNA mutations eventually cause irreplaceable cell losses through programmed cellular death. One of the biggest surprises from this work is that premature aging in mutator mice appears to occur without a major increase or spike in levels of reactive oxygen species (ROS) or oxidative stress. These data, along with other clues in the literature, have resulted in a fresh look at the fundamental role of reactive oxygen species in aging and age-dependent diseases, a topic discussed in detail in Chapters 11 to 13.

Mutator mice have very high levels of single base mutations (point mutations) as well as high levels of deletions, the latter of which actually shorten the circumference of the mtDNA circle. This is an important advance in the field of aging because it establishes that point mutations among the 13 protein-coding and energy-related genes cause premature aging (Edgar et al., 2009). The current model for premature aging in mutator mice proposed by these researchers is that accumulation of point mutations of mtDNA leads to dysfunctional respiratory chain subunits caused by single amino acid substitutions along the peptide chain. These defects are proposed to destabilize respiratory chain complexes, slowing energy production until a critical threshold is reached, which in the mouse model of aging triggers widespread cellular death.

Mutations in another gene, the *cisd2* gene in mice, with homologs in humans, have recently been shown to accelerate aging in mice as well as humans (Chen et al., 2009). In knockout mutants of mice, disruption of CISD2, a redox-active iron-sulfur protein targeted to mitochondria, leads to accelerated aging. Knockout mice display thinner bones and hair, corneal opacities and degeneration, decreased muscle mass, prominent eyes, and protruding ears, all of which are consistent with premature aging. Mitochondria isolated from the mutant mice show a defect in respiration, an indication that ATP production is slowed.

9.3 MUTATOR GENES ACCELERATE NEURODEGENERATION IN HUMANS, SUGGESTING THAT THE BRAIN CAN SET THE PACE OF AGING

There are no genetically engineered mutator people walking around in which human DNA polymerase gamma is purposely altered to accelerate rates of mtDNA mutations such as seen in mutator mice. However, the replication fidelity of human mitochondrial DNA polymerase has been found to be altered by naturally occurring mutations found infrequently in the general population (Chapter 8). Indeed, mutations in the catalytic subunit of DNA polymerase (POLG) are believed to create the same kind of mutator effect as seen in mice, increasing rates of mutation in mtDNA and lowering cellular energy levels. Other genes supporting mtDNA replication and repair also exhibit a mutator effect when humans acquire these mutated genes. And as introduced in Chapter 8, each parent harboring a different mutator mutation can result in offspring carrying double mutations in mitochondrial DNA polymerase. A wide spectrum of disease states in humans arises as the result of naturally occurring DNA polymerase gamma, mutations creating pathologies, as follows (also see Chapter 8):

- Parkinson's disease
- Breast cancer
- Ophthalmoplegia (progressive external opthalmoplegia [PEO])
- Apert syndrome
- Neuropathy
- Dysarthria
- Sensory ataxic neuropathy, dysarthria, and ophthalmoparesis (SANDO)
- Male infertility

These and other human diseases and symptoms attributed to the mutator properties of mtDNA polymerase are reviewed by several authors (Longley et al., 2005; Chan and Copeland, 2009; Van Goethem et al., 2003; Singh et al., 2009; Hudson and Chinnery, 2006; Luoma et al., 2004; Invernizzi et al., 2008; Turnbull et al., 2010; Taylor and Turnbull, 2005; Seo et al., 2010; Hudson et al., 2008; Bensch et al., 2009). These new findings are stimulating research in the fields of human aging, dementia, and cancer and reinforce the critical roles played by mitochondria/energy stress in aging and age-related diseases.

A recent comprehensive analysis of POLG-mediated human diseases shows that mutations of POLG often cause neurodegeneration (Wong et al., 2008). These data bring the fields of aging, neurodegenerative diseases, and membranes closer together. The key point is that many of these phenotypes presumably caused by energy stress have been traced to premature death of neurons in the brain or peripheral nervous system. Obviously, membranes of neurons are unique compared to most human cells in being enriched with DHA, the most readily peroxidized fatty acid in nature. Do DHA membranes hold special power over aging in neurons (see Valentine and Valentine, 2013)?

9.4 RECENT DATA CONFIRM THE MUTATOR CONCEPT AND HELP EXPLAIN HOW MITOCHONDRIAL FUSION CAN PUSH BACK AGAINST AGING

Mitochondria often fuse together, forming novel hybrids (Nakada et al., 2009; Chen et al., 2009; Chan, 2006; Chen and Chan, 2006) (Figure 9.2). Fusion allows aging mitochondria to continually exchange contents, including mutationally altered mtDNA. These data derived from studies of the molecular biology of mitochondrial fusion in mice show that fusional exchange somehow preserves mtDNA function in the face of mutations (Chen et al., 2010). Mitochondrial fusion safeguards both mtDNA integrity and copy number. This pioneering work opens a new chapter in the story of the mitochondrial hypothesis of aging in which exchange of contents of aging mitochondria likely acts to protect cells against aging and age-related diseases.

In mammalian cells, three large enzymes called GTPases are important for mitochondrial fusion, which involves the coordinated fusion of the outer and inner mitochondrial membranes. The mitofusins Mfn1 and Mfn2 are located on the mitochondrial outer membrane and catalyze early steps in membrane fusion (Song et al., 2009). The dynamin-related protein OPA1 is localized on the inner membrane and

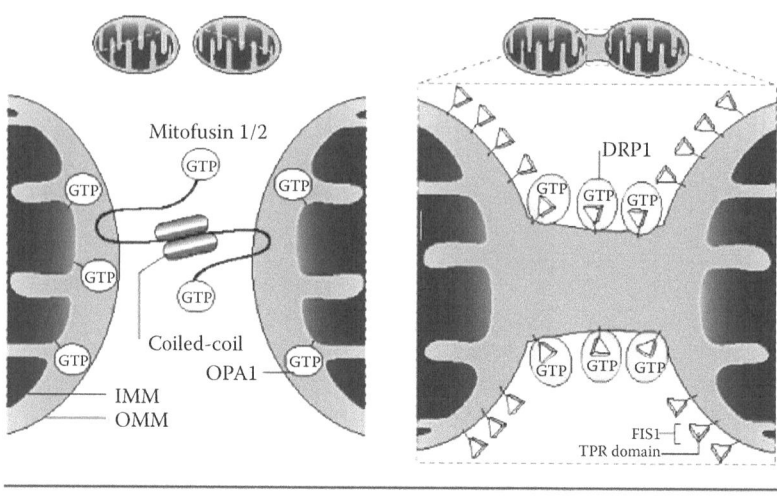

FIGURE 9.2 Mitochondrial fusion protects cells against energy stress caused by accumulation of mutations in mtDNA. The left half of the diagram shows two mitochondria being tethered together by mitofusin complexes. Mfn1 is localized to the mitochondrial inner membrane with both its amino- and carboxyl-terminal regions exposed to the cytosol. Homotypic interactions between Mfn1 molecules span adjacent mitochondria and are mediated by the heptad region HR2 (cylinders). OPA1, a dynamin family GTPase, resides in the intermembrane space in association with the inner membrane. Energy for tethering is supplied by GTP. Mitochondrial fission shown in the right panel is discussed in Chapter 17. (From Youle and Karbowski, *Nat. Rev. Mol. Cell Biol.* 6:657–63, 2005. Copyright © 2005. Reprinted by permission from Macmillan Publishers Ltd.)

is essential for inner membrane fusion (Song et al., 2009). Mutations in Mfn2 and OPA1 cause two neurodegenerative diseases—Charcot-Marie-Tooth type 2Q and dominant optic atrophy, respectively. Thus, mitochondrial fusion clearly protects mitochondrial function (Detmer and Chan, 2007a). Studies of numerous defects seen in mitochondria missing the mitofusin Mfn1 have led to a model in which mitochondrial fusion protects mitochondrial function by enabling content mixing (Detmer and Chan, 2007b). Because of content mixing, it has been hypothesized that mitochondrial fusion may be involved in the ability of human cells to tolerate high levels of pathogenic mtDNA, such as generated during aging (Nakada et al., 2009).

A signature feature of mitochondrial genetics is that most mtDNA mutations that accumulate with aging are compensated by nonmutated genes on other circles of mtDNA present in the same mitochondrion. The net effect is that mutations in mitochondria must accumulate to high levels before effects on respiration and oxidative phosphorylation (i.e., energy stress) can be detected (Taylor and Turnball, 2005). For example, cells containing both mutant and wild-type mtDNA can accumulate up to 60 to 90 percent mutationally pathogenic mtDNA molecules without a noticeable decline in respiratory activity (Rossignol et al., 2003). This threshold effect has important implications for aging because it predicts that energy stress caused by accumulation of mutations will not emerge until many mtDNA mutations have accumulated. Recent data using knockout mutations of mitofusins in mice show a dramatic effect of mitochondrial fusion in mtDNA replication fidelity and stability (Chen et al., 2010). For example, muscle cells from seven- to eight-week-old mice missing both Mfn1 and Mfn2 contain only about 250 copies of mtDNA, compared to 3500 in control animals.

Studies of rates of accumulation of mutations in mtDNA in double-mitofusin knockout mice show a remarkable trend. Muscle from seven- to eight-week-old mice harbors a five-fold increase in mtDNA point mutations and a fourteen-fold increase in deletions compared to wild-type mice. Older animals from eight to thirteen months old carrying mutations in either of the mitofusin genes Mfn1 or Mfn2 exhibit an eighty-fold increase in mtDNA deletions. This extraordinary age-dependent increase in levels of deletions did not occur with point mutations, suggesting that the mechanism causing mtDNA deletions is more sensitive to a decrease in mitochondrial fusion.

Mutator mice, as discussed above, produce a mutant form of mtDNA polymerase with a deficient proofreading domain, causing cells to accumulate mutations at an accelerated rate. Combining mutations in POLGA and Mfn1 leads to a synergistic effect that is neonatal lethal (Chen et al., 2010). In contrast, mice carrying either one of the genes survive well into adulthood. Having ruled out an increased mutation rate as responsible for the lethal effect in the double mutant, biochemical tests were devised to measure bioenergetic parameters at the cellular level. Mouse embryonic fibroblasts were used in these experiments. These data show that oxygen consumption is lowered by 95 percent in cells harboring the double mutation, and ATP levels drop more than 98 percent compared to wild-type cells. These data clearly demonstrate that these mutations cause energy stress in these cells and likely force them to use alternative bioenergetic mechanisms such as glycolysis. The most profound block in the respiratory chain occurred at complex 1, resulting in a twenty-fold reduction in activity. Recall that seven different subunits of complex 1 are encoded by mtDNA, making genes of complex 1 the largest target for mutations altering energy-transducing machinery coded by mitochondrial DNA. These data are consistent with and further illuminate the revised mitochondrial hypothesis of aging, in which errors of mtDNA replication generate mutations in mtDNA, causing energy stress and aging (Larsson, 2010).

9.5   MIXING OF COMPONENTS DURING MITOCHONDRIAL FUSION MIGHT PROTECT MEMBRANES AGAINST AGE-DEPENDENT DAMAGE

Mitochondrial fusion is now viewed as a universal mechanism to preserve and protect mitochondria. There is an interesting analogy in the world of plant breeding in which crossing two plants, each exhibiting mediocre agronomic properties, creates a plant more vigorous than either parent plant, a phenomenon called hybrid vigor. Fusion of mitochondria creates a miniaturized form of hybrid vigor, in which the hybrid mitochondria are healthier than the parents. Individual mitochondria contain multiple copies of mtDNA, and during fusion, exchange of mtDNA takes place, fostering both biochemical and genetic complementation. Mitochondrial fusion has some earmarks of a novel form of bacterial sexuality, in which benefits of DNA exchange define the sexual process. We are not saying that mtDNA recombination is occurring or is responsible; rather, genetic complementation seems to account for these results. Considerable attention is now focused on how swapping mtDNA or other mitochondrial components during fusion preserves and protects mitochondria during aging.

It is well known that during bacterial sex a DNA strand originating from the donor cell is transferred to the recipient cell, with both plasmids and chromosomal DNA mobilized by this mechanism. In contrast, so-called sexuality of mitochondria involves not only the two-way exchange of mtDNA, but also exchange of other biochemical components localized in the space between outer and inner membranes, as well as components present in the mitochondrial matrix. Thus, a great number of different components appear to be exchanged during mitochondrial fusion. The focus next is on exchange of components that might protect membranes and enable longevity. A selected list of possible membrane protective molecules is as follows:

- Phospholipids or their precursors needed to repair or dilute out damaged membranes
- Antioxidants (both water soluble and lipophilic)
- Membrane repair lipases
- ROS detoxification enzymes
- Cardiolipin synthesis enzymes

This list continues, and further research is needed to determine if any of these exchanges enhance membrane stability.

The final point deals with other possible mechanisms to explain the large decrease in bioenergetic capacity of mitochondria carrying mutations of mitofusins or combinations of mutations such as mitofusins and POLG. Detmer and Chan (2007a) show that mitochondria of fusion-minus mice undergo dramatic miniaturization. Wild-type mitochondria as long as ten microns were downsized to one-tenth this length. We suggest that miniaturization alone might affect bioenergetic properties and the fate of mitochondria. Hidden benefits may accompany the increased surface-to-volume ratio of miniaturized mitochondria. However, there are some ominous signs of risk associated with the highly unsaturated nature of the membranes of these mini-mitochondria (Kirkland et al., 2002). For example, the high surface-to-volume ratio of mini-mitochondria that favors higher rates of O_2 diffusion simultaneously exposes mitochondrial membranes to oxidative damage, perhaps increasing leakage of protons causing energy uncoupling. Whether mini-mitochondrial membranes are subjected to increased leakage of protons caused by curvature stress remains an interesting question. Other concerns are that small mitochondria may have limited capacity to dock to and exchange lipids with the endoplasmic reticulum (ER) during fission or may display dysfunctional transport properties to and from distant regions of large cells, especially axons (see Chapter 17). Also, the extraordinary increase in numbers of defective mini-mitochondria might cause accumulation of toxic mito-chondria, overwhelming mitophagy machinery (Chapter 18). Changes in ROS pro-duction by mini-mitochondria remain an open possibility.

9.6   SUMMARY

Discoveries using mutator mice show that mutations accumulating in mtDNA as the result of errors of replication by mtDNA polymerase eventually decrease energy pro-duction in mitochondria, causing aging. A similar mutator gene occurring naturally in humans has been found to cause a variety of neurodegenerative diseases, likely by a mechanism similar to generalized aging in mice. Mitochondrial fusion mediated by mitofusins has been found to act as a mechanism to protect cells against energy stress caused by accumulation of mutations of mtDNA during aging. These data confirm and extend earlier results using mutator mice and open a new chapter in the story of human aging. These data also suggest that the conventional oxidative stress theory of aging requires revision. However, it is premature to suggest that ROS don't play important, if indirect, roles in aging (Chapters 11 through 13).

REFERENCES

Amati-Bonneau, P., M. L. Valentino, P. Reynier, et al. 2008. OPA1 mutations induce mito-chondrial DNA instability and optic atrophy 'plus' phenotypes. *Brain* 131:338–51.

Bensch, K. G., J. L. Mott, S. W. Chang, et al. 2009. Selective mtDNA mutation accumulation results in beta-cell apoptosis and diabetes development. *Am. J. Physiol. Endocrinol. Metab.* 296:E672–80.

Chan, D. C. 2006. Mitochondrial fusion and fission in mammals. *Annu. Rev. Cell Dev. Biol.* 22:79–99.

Chan, S. S., and W. C. Copeland. 2009. DNA polymerase gamma and mitochondrial dis-ease: understanding the consequence of POLG mutations. *Biochim. Biophys. Acta* 1787:312–19.

Chen, H., and D. C. Chan. 2006. Critical dependence of neurons on mitochondrial dynamics. *Curr. Opin. Cell Biol.* 18:453–59.

Chen, H., M. Vermulst, Y. E. Wang, et al. 2010. Mitochondrial fusion is required for mtDNA stability in skeletal muscle and tolerance of mtDNA mutations. *Cell* 141:280–89.

Chen, Y. F., C. H. Kao, Y. T. Chen, et al. 2009. Cisd2 deficiency drives premature aging and causes mitochondria-mediated defects in mice. *Genes Dev.* 23:1183–94.

Detmer, S. A., and D. C. Chan. 2007a. Functions and dysfunctions of mitochondrial dynamics. *Nat. Rev. Mol. Cell. Biol.* 8:870–79.

Detmer, S. A., and D. C. Chan. 2007b. Complementation between mouse Mfn1 and Mfn2 pro-tects mitochondrial fusion defects caused by CMT2A disease mutations. *J. Cell. Biol.* 176:405–14.

Dogan, S. A., and A. Trifunovic. 2011. Modelling mitochondrial dysfunction in mice. *Physiol. Res.* 60(Suppl 1):S61–70.

Edgar, D., I. Shabalina, Y. Camara, et al. 2009. Random point mutations with major effects on protein-coding genes are the driving force behind premature aging in mtDNA mutator mice. *Cell Metab.* 10:131–38.

Edgar, D., and A. Trifunovic. 2009. The mtDNA mutator mouse: dissecting mitochondrial involvement in aging. *Aging* 1:1028–32.

Efremov, R. G., R. Baradaran, and L. A. Sazanov. 2010. The architecture of respiratory complex I. *Nature* 465:441–45.

Hudson, G., P. Amati-Bonneau, E. L. Blakely, et al. 2008. Mutation of OPA1 causes dominant optic atrophy with external ophthalmoplegia, ataxia, deafness and multiple mitochondrial DNA deletions: a novel disorder of mtDNA maintenance. *Brain* 131:329–37.

Hudson, G., and P. F. Chinnery. 2006. Mitochondrial DNA polymerase-gamma and human disease. *Hum. Mol. Genet.* 15(spec. no. 2):R244–52.

Invernizzi, F., S. Varanese, A. Thomas, et al. 2008. Two novel POLG1 mutations in a patient with progressive external ophthalmoplegia, levodopa-responsive pseudo-orthostatic tremor and parkinsonism. *Neuromuscl. Disord.* 18:460–64.

Kirkland, R. A., R. M. Adibhatla, J. F. Hatcher, et al. 2002. Loss of cardiolipin and mitochondria during programmed neuronal death: evidence of a role for lipid peroxidation and autophagy. *Neuroscience* 115:587–602.

Kukat, A., D. Edgar, I. Bratic, et al. 2011. Random mtDNA mutations modulate proliferation capacity in mouse embryonic fibroblasts. *Biochem. Biophys. Res. Commun.* 409:394–99.

Larsson, N. G. 2010. Somatic mitochondrial DNA mutations in mammalian aging. *Annu. Rev. Biochem.* 79:683–706.

Longley, M. J., M. A. Graziewicz, R. J. Bienstock, et al. 2005. Consequences of mutations in human DNA polymerase gamma. *Gene* 354:125–31.

Luoma, P., A. Melberg, J. O. Rinne, et al. 2004. Parkinsonism, premature menopause, and mitochondrial DNA polymerase gamma mutations: clinical and molecular genetic study. *Lancet* 364:875–82.

Nakada, K., A. Sato, and J. Hayashi. 2009. Mitochondrial functional complementation in mitochondrial DNA-based diseases. *Int. J. Biochem. Cell Biol.* 41:1907–13.

Niu, X., A. Trifunovic, N. G. Larsson, et al. 2007. Somatic mtDNA mutations cause progressive hearing loss in the mouse. *Exp. Cell Res.* 313:3924–34.

Paradies, G., G. Petrosillo, V. Paradies, et al. 2011. Mitochondrial dysfunction in brain aging: role of oxidative stress and cardiolipin. *Neurochem. Int.* 58:447–57.

Rossignol, R., B. Faustin, C. Rocher, et al. 2003. Mitochondrial threshold effects. *Biochem. J.* 370:751–62.

Seo, A. Y., A. M. Joseph, D. Dutta, et al. 2010. New insights into the role of mitochondria in aging: mitochondrial dynamics and more. *J. Cell Sci.* 123:2533–42.

Singh, K. K., V. Ayyasamy, K. M. Owens, et al. 2009. Mutations in mitochondrial DNA polymerase-gamma promote breast tumorigenesis. *J. Hum. Genet.* 54:516–24.

Someya, S., T. Yamasoba, G. C. Kujoth, et al. 2008. The role of mtDNA mutations in the pathogenesis of age-related hearing loss in mice carrying a mutator DNA polymerase gamma. *Neurobiol. Aging* 29:1080–92.

Song, Z., M. Ghochani, J. M. McCaffery, et al. 2009. Mitofusins and OPA1 mediate sequential steps in mitochondrial membrane fusion. *Mol. Biol. Cell.* 20:3525–32.

Taylor, R. W., and D. M. Turnbull. 2005. Mitochondrial DNA mutations in human disease. *Nat. Rev. Genet.* 6:389–402.

Trifunovic, A., and N. G. Larsson. 2008. Mitochondrial dysfunction as a cause of ageing. *J. Intern. Med.* 263:167–78.

Trifunovic, A., A. Wredenberg, M. Falkenberg, et al. 2004. Premature ageing in mice expressing defective mitochondrial DNA polymerase. *Nature* 429:417–23.

Turnbull, H. E., N. Z. Lax, D. Diodato, et al. 2010. The mitochondrial brain: from mitochondrial genome to neurodegeneration. *Biochim. Biophys. Acta* 1802:111–21.

Tyynismaa, H., and A. Suomalainen. 2009. Mouse models of mitochondrial DNA defects and their relevance for human disease. *EMBO Rep.* 10:137–43.

Valentine, R. C., and D. L. Valentine. 2013. *Neurons and the DHA principle*. Boca Raton, FL: Taylor and Francis Group.

Van Goethem, G., J. J. Martin, B. Dermaut, et al. 2003. Recessive POLG mutations presenting with sensory and ataxic neuropathy in compound heterozygote patients with progressive external ophthalmoplegia. *Neuromuscul. Disord.* 13:133–42.

Wong, L. J., R. K. Naviaux, N. Brunetti-Pierri, et al. 2008. Molecular and clinical genetics of mitochondrial diseases due to POLG mutations. *Hum. Mutat.* 29:E150–72.

Youle, R. J., and M. Karbowski. 2005. Mitochondrial fission in apoptosis. *Nat. Rev. Mol. Cell Biol.* 6:657–63.

10 Benefits of Polyunsaturated Mitochondrial Membranes

The benefits of mitochondrial membrane unsaturation are highlighted here. As background, scientists in the 1970s were first able to calibrate the physical speed of the essential light-sensing protein rhodopsin embedded in docosahexaenoic acid (DHA)-enriched membrane disks (i.e., rhodopsin disks) of rod cells of the eye (Poo and Cone, 1974; Liebman and Entine, 1974; Litman et al., 2001). These now classic studies demonstrate that the light-sensing protein rhodopsin rotates and moves laterally across the surface of membrane disks at amazing speeds. The motional properties of the integral membrane protein rhodopsin were calibrated using micro-lasers; for example, rhodopsin was found to rotate in a molecular swirl timed at 0.00002 s per turn. This remains the speed record for a full swing of a membrane-bound protein and links highly unsaturated phospholipids with extreme motion of membrane components. Lateral motion of rhodopsin across the surface of membrane disks is also extremely fast and is essential in triggering the visual cascade. By analogy, we suggest that the unique conformational dynamics of polyunsaturated fatty acids (PUFAs) in mitochondrial membranes are harnessed to maximize motion of components of the electron transport chain, resulting in enhanced energy production. We define motion of mitochondrial membranes in terms of mobility of membrane lipids, proteins, and the lipophilic electron carrier ubiquinone—properties that enable energy production.

10.1 CONFORMATIONAL DYNAMICS EXPLAIN HOW POLYUNSATURATED CHAINS ARE HARNESSED FOR BIOENERGETIC GAIN

Because the DHA tails of membrane phospholipids are in perpetual motion (Feller, 2008a, 2008b; Feller et al., 2002, 2003; Feller and Gawrisch, 2005; Stillwell and Wassall, 2003; Soubias and Gawrisch, 2007), these chains don't stand still long enough (Figure 10.1) to bind with their neighbors and harden as for, e.g., butter or lard. Thus, the contortions of highly unsaturated membrane fatty acids such as DHA chains keep the membrane surface in constant motion even in mitochondria of ectothermic organisms living in the extreme cold (Valentine and Valentine, 2009).

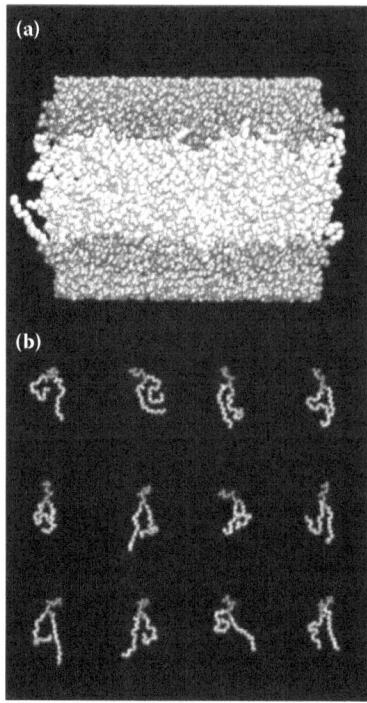

FIGURE 10.1   See color insert. Extraordinary conformational dynamics of DHA. (a) DHA-enriched membrane. (b) Dynamic conformations of DHA phospholipids. (Images courtesy of Scott Feller, and generated by Matthew B. Roark, both of Wabash College.)

DHA contributes extraordinary motion to membrane components (e.g., spinning and lateral movement), as seen with rhodopsin protein in rhodopsin membrane disks. Chemists have found that DHA oils resist hardening until temperatures reach lower than −50°C (Koynova and Caffrey, 1998). This property is important in nature, allowing mitochondria in cold-adapted animals to function at near-freezing temperatures. In humans molecular species of cardiolipin with four 18:2 chains are common, and the conformational dynamics of $(18:2)_4$-CL (Lewis and McElhaney, 2009) may enhance energy production.

The cytochromes of the respiratory chain in mammals have been calibrated to move rapidly in polyunsaturated mitochondrial membranes, though at significantly lower rates than rhodopsin protein (Dixit and Vanderkooi, 1984). Other studies show that collisions among lipid-soluble electron carriers, ubiquinone in mitochondria (Gupte et al., 1984) and plastoquinone in chloroplasts (Blackwell et al., 1994), are often rate-limiting steps in electron transport chains of these organelles. Ubiquinone is hindered as it shuttles electrons among the first three enzymes of the respiratory chain (complexes 1–3). Obstruction is due to the high densities of proteins populating the membrane surface as well as lipid rafts. In essence, the distance traveled by ubiquinone through the membrane is lengthened by obstacles, including respiratory complexes and lipid rafts.

10.2 OVEREXPRESSING DHA/EPA IN TRANSGENIC MICE INCREASES RATES OF ELECTRON TRANSPORT

Hagopian and colleagues (2010) recently used isolated mitochondria from transgenic mice overexpressing EPA (2.55 percent) and DHA (1.74 percent) to test effects of omega-3 on electron transport activity and reactive oxygen species (ROS) production. The above numbers in parentheses refer to increases in levels of DHA/EPA calculated as percent of total fatty acids of mitochondrial membranes in recombinant versus wild-type mice. We propose combining the numbers for EPA and DHA, giving a total increase of 4.29 percent highly unsaturated fatty acids (HUFAs) in mitochondrial membranes of transgenic *fat*-1 mice compared to wild-type controls. This value is roughly equivalent to raising the incubation temperature for measuring electron transport by 8.5°C (see Valentine and Valentine, 2009). Most of the increase in DHA plus EPA is found in phosphatidylethanolamine, with a lesser amount in phosphatidylserine; no significant changes were found in cardiolipin fatty acid composition. Enriching mitochondrial membranes with 4.29 percent DHA plus EPA is predicted to significantly alter the lipid dynamics and rates of collisions among components of the electron transport chain. Hagopian and coworkers (2010) carried out an extensive analysis of rates of electron transport complexes in isolated mitochondria enriched with DHA plus EPA (Figure 10.2a) compared to controls. The most striking change was seen with complex 3 (Figure 10.2b), where activities more than doubled from about 600 units in controls to about 1300 units in mitochondria from transgenic mice. This powerful effect of HUFAs on activity of complex 3 was enough to increase the coupled flow of electrons from complex 1 through complex 3. These data are consistent with the collision model of electron transport discussed above and in a previous book (Valentine and Valentine, 2009). Further interpretation of the data of Hagopian and colleagues (2010) on mitochondrial function is covered in Chapter 11.

We suggest that the evolution in humans of highly polyunsaturated molecular species of cardiolipin such as $(18:2)_4$-CL is another mechanism to improve motion in mitochondrial membranes. We propose that the four 18:2 tails of CL generate a highly dynamic structure (Lewis and McElhaney, 2009) suggested to maximize rates of collisions between ubiquinone and its redox partners. In view of data by Hagopian and colleagues discussed above, extraordinary membrane motion enhancing collisions between ubiquinone and complex 3 might overcome a bottleneck in the electron transport chain.

10.3 MITOCHONDRIAL MEMBRANE COMPOSITION REFLECTS THE NEED TO BALANCE ENERGY PRODUCTION AND ENERGY CONSERVATION

Mitochondrial membranes are responsible for both energy production and proton fidelity (Figure 10.3). The term *proton fidelity* is used to describe the permeability properties of mitochondrial and other membranes against spontaneous escape or leakage of protons across the lipid portion of the membrane. High proton fidelity refers

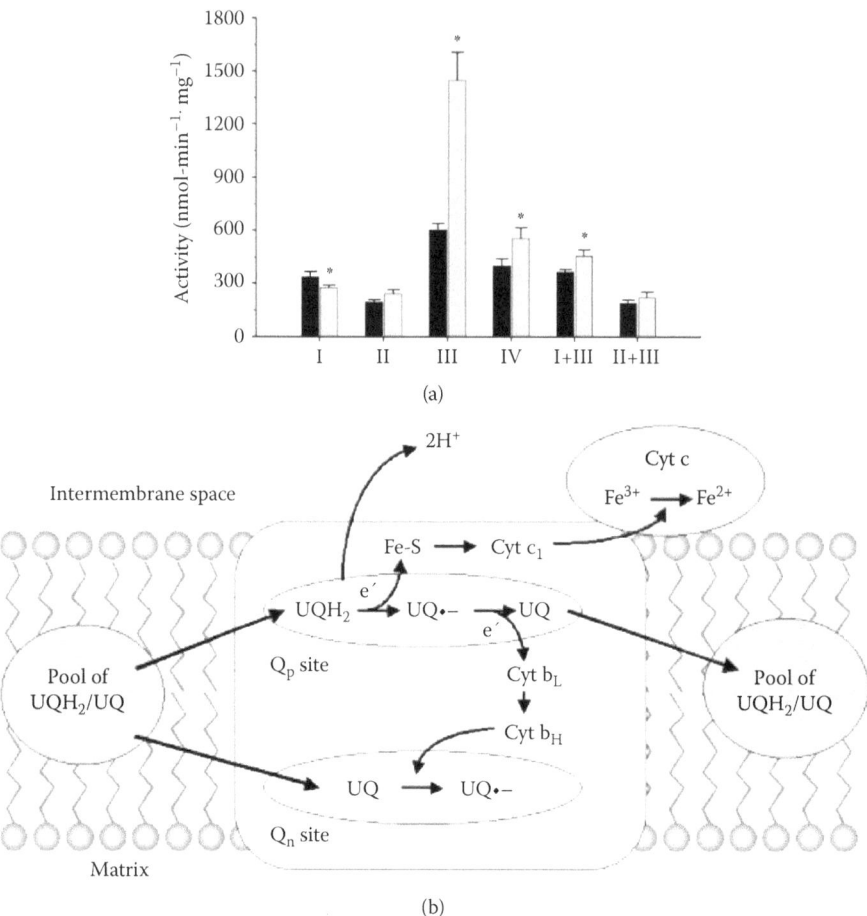

(a)

(b)

FIGURE 10.2 Overexpression of DHA/EPA in recombinant mice results in enrichment of mitochondrial membranes and higher rates of electron transport. (a) Gray bars show enhanced rates of electron transport in mitochondria enriched with DHA/EPA. Note the more than doubling of activity of complex 3 in DHA/EPA-enriched mitochondria compared to mitochondria from controls (black bars). (From Hagopian et al., *PLoS One* 5(9):e12696, 2010.) (b) Mode of action of complex 3 requires rapid movement of the lipid-soluble ubiquinone electron carriers UQ and UQH_2 within the bilayer as they shuttle electrons between sites of complex 3. We suggest that enrichment of mitochondrial membranes with DHA/EPA speeds up collisions between UQH_2/UQ and the sites they service on complex 3. This diagram shows only the first half of the Q-cycle of electron flow catalyzed by complex 3. UQ and UQH_2 stand for oxidized ubiquinone and fully reduced ubiquinone, respectively. (Courtesy of Dr. Michael Blaber, Florida State University.)

to membranes whose molecular architecture is robust or tight against proton tunneling. Proton tunneling is considered to be a fundamental property of all biological membranes. Protons are monovalent cations that follow established chemical principles, tending to spontaneously migrate from regions of high concentration in cellular cytoplasm to low concentrations in the mitochondrial matrix. Proton concentrations

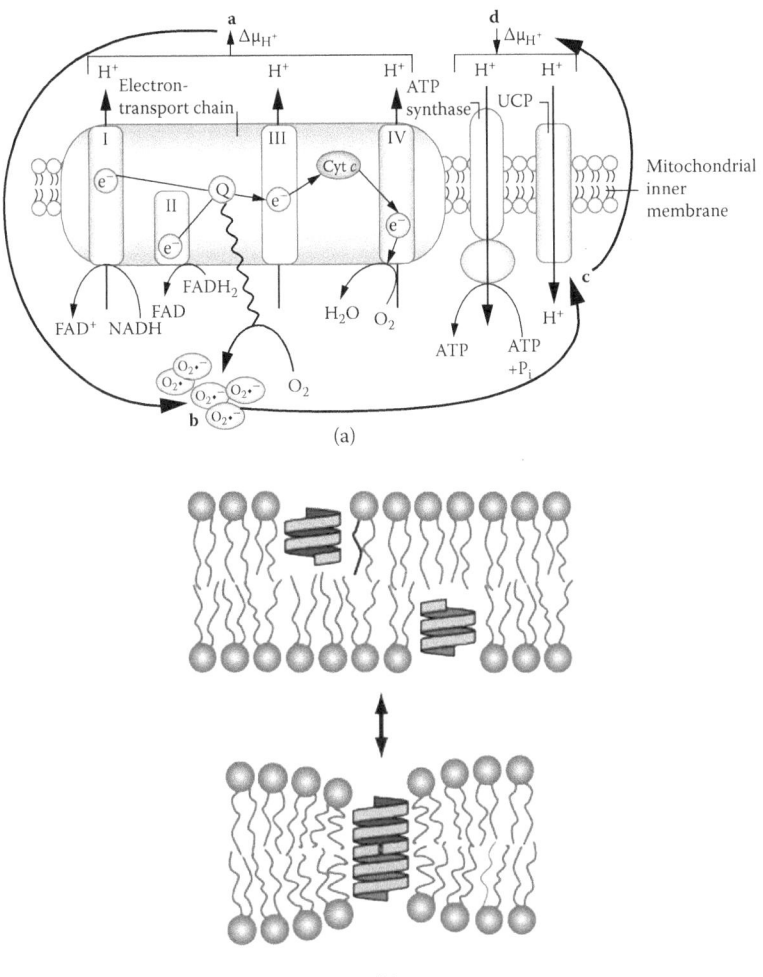

(a)

(b)

FIGURE 10.3 Overview of two mechanisms of proton leakage across the inner mitochondrial membrane. (a) The first mechanism involves proton leaks catalyzed by uncoupling proteins (UCPs) (also see Chapter 16). (From Krauss et al., *Nat. Rev. Mol. Cell. Biol.* 6:248–61, 2005. Copyright © 2005. Reprinted by permission from Macmillan Publishers Ltd.) (b) The second mechanism involves direct movement of protons across the phospholipid portion of the membrane, presumably via a water-wire mechanism similar to that of gramicidin, as shown in this diagram. Two gramicidin monomers join together to span the lipid bilayer. The core of gramicidin forms a channel holding 21 water molecules in single file. Protons move at amazing speed across this water-wire, uncoupling proton electrochemical gradients of cells. Water-wires forming spontaneously in membranes, including mitochondrial membranes, are believed to function by a similar mechanism. The role of water-wires in bacterial membranes is clearly established, but the importance of lipid-mediated proton tunneling in mitochondrial membranes has not been settled. (From Khandelia et al., *Biochim. Biophys. Acta* 1778:1528–36, 2008. Copyright © 2008. Reprinted with permission from Elsevier.)

outside mitochondria are isolated by the inner mitochondrial membrane, which acts as a permeability barrier. Thus, the molecular architecture of lipids of the inner membrane governs proton fidelity of the membrane (Figure 10.3a and b). Several properties that modulate proton fidelity of the lipid portion of membranes are as follows:

- Membrane thickness
- Unsaturation levels where each double bond thins membranes by roughly one C-C bond length
- Temperature
- Pressure
- Bulky groups (e.g., additional methyl groups added to fatty acid chains)
- Cholesterol
- Plasmalogens (?)
- Chaotropic agents such as urea
- Solvents such as toluene

In addition to these parameters, switching from proton to sodium bioenergetics is a common mechanism used by archaea and bacteria to grow under conditions of energy stress or chronic proton uncoupling. The selective advantage is that Na^+ is about three orders of magnitude less permeable than H^+ (see Chapter 4). Numerous other mechanisms likely govern proton fidelity, including the evolution by archaea of robust isoprenoid membrane monolayers, which set the standard for proton and sodium fidelity among prokaryotes. Archaeal membranes are several orders of magnitude less permeable to spontaneous leakage of protons and sodium than bacteria. Thus, archaeal membranes exhibit by far the highest proton fidelity, with mitochondrial membranes expected to display relatively low proton fidelity by comparison (Valentine, 2007).

Over the past two decades a number of papers exploring mitochondrial proton leaks have led to opposing conclusions (Brand et al., 1994; Porter and Brand, 1993; Stillwell et al., 1997). Some data are consistent with a close linkage between mitochondrial membrane unsaturation levels and proton leakage, with increasing unsaturation observed to accelerate membrane lipid-mediated proton leaks (in contrast to proton leakage mediated by membrane proteins). These data also suggest that mitochondrial proton leakage contributes significantly to energy stress. Stillwell and colleagues (1997) measured the effects of DHA levels on proton leaks of mitochondria of mice. These researchers used several different procedures to manipulate DHA levels in mitochondrial membranes of mice with the conclusion that increasing DHA levels in mitochondrial phospholipids results in increased proton leaks, and vise versa. Increased DHA levels were also found to increase membrane fluidity while decreasing mitochondrial membrane potentials (Stillwell et al., 1997). These data are consistent with DHA playing both beneficial and harmful roles in mitochondrial membranes of mice. DHA knockout mutants of mice are now available in which DHA levels in liver tissue, and presumably mitochondrial membranes, plummet sharply. It would be of interest to compare the H^+ leak rates of mitochondria of DHA knockout mutants of mice with wild-type mitochondria. Our prediction is that rates of H^+ leaks in mitochondria isolated from DHA knockout mice, in which DHA might be largely replaced by 18:2, would be significantly lower than those from wild-type mice with high DHA levels.

In contrast to the above data, Jastroch and colleagues (2010) suggest that proton leakage across the lipid portion of mitochondrial membranes plays only a minor role compared to protein-assisted uncoupling of proton gradients (Figure 10.3a; see Chapter 16). However, these data do not rule out the possibility that under certain conditions spontaneous proton leakage mediated across the lipid portion of the membrane might contribute significantly to mitochondrial energy status.

10.4 SUMMARY

The high levels of PUFAs in human mitochondrial membranes are proposed to be highly beneficial, helping maximize rates of electron transport and energy production. The same unique chemistry of PUFAs that facilitates rapid collisions among components of the respiratory chain and enhances energy production is also proposed to increase the probability that a lethal peroxidation-mediated chain reaction might be propagated by oxidatively damaged mitochondria. This concept is discussed in more detail in Chapters 11 to 13. Thus, human mitochondria in the evolution of their PUFA-enriched inner mitochondrial membrane likely faced a delicate balancing act featuring great benefits in terms of bioenergetic gain versus risks caused by oxidative and energy stress. We suggest that the dualities of polyunsaturated fatty acid chains help explain the roles of mitochondrial membranes during aging.

REFERENCES

Blackwell, M., C. Gibas, S. Gygax, D. Roman, and B. Wagner. 1994. The plastoquinone diffusion coefficient in chloroplasts and its mechanistic implications. *Biochim. Biophys. Acta* 1183:533–43.

Brand, M. D., L. F. Chien, E. K. Ainscow, et al. 1994. The causes and functions of mitochondrial proton leak. *Biochim. Biophys. Acta* 1187:132–39.

Dixit, B. P. S. N., and J. M. Vanderkooi. 1984. Probing structure and motion of the mitochondrial cytochromes. *Curr. Top. Bioenerg.* 13:159–202.

Feller, S. E. 2008a. Lipid animation: 500 ps dynamics of a stearoyl docosohexaenoyl PC lipid molecule (14MB). Wabash College Chemistry Department, Crawfordville, IN. http://persweb.wabash.edu/facstaff/fellers/animations.html.

Feller, S. E. 2008b. Acyl chain conformations in phospholipid bilayers: a comparative study of docosahexaenoic acid and saturated fatty acids. *Chem. Phys. Lipids* 153:76–80.

Feller, S. E., and K. Gawrisch. 2005. Properties of docosahexaenoic acid-containing lipids and their influence on the function of rhodopsin. *Curr. Opin. Struct. Biol.* 15:416–22.

Feller, S. E., K. Gawrisch, and A. D. MacKerell Jr. 2002. Polyunsaturated fatty acids in lipid bilayers: intrinsic and environmental contributions to their unique physical properties. *J. Am. Chem. Soc.* 124:318–26.

Feller, S. E., K. Gawrisch, and T. B. Woolf. 2003. Rhodopsin exhibits a preference for solvation by polyunsaturated docosahexaenoic acid. *J. Am. Chem. Soc.* 125:4434–35.

Gupte, S., E. S. Wu, L. Hoechli, et al. 1984. Relationship between lateral diffusion, collision frequency, and electron-transfer of mitochondrial inner membrane oxidation-reduction components. *Proc. Natl. Acad. Sci. USA* 81:2606–10.

Hagopian, K., K. L. Weber, D. T. Hwee, et al. 2010. Complex I-associated hydrogen peroxide production is decreased and electron transport chain enzyme activities are altered in n-3 enriched fat-1 mice. *PLoS One* 5(9):e12696.

Jastroch, M., A. S. Divakaruni, S. Mookerjee, et al. 2010. Mitochondrial proton and electron leaks. *Essays Biochem.* 47:53–67.

Khandelia, H., J. H. Ipsen, and O. G. Mouritsen. 2008. The impact of peptides on lipid membranes. *Biochim. Biophys. Acta* 1778:1528–36.

Koynova, R., and M. Caffrey. 1998. Phases and phase transitions of the phosphatidylcholines. *Biochim. Biophys. Acta* 1376:91–145.

Krauss, S., C. Y. Zhang, and B. B. Lowell. 2005. The mitochondrial uncoupling-protein homologues. *Nat. Rev. Mol. Cell. Biol.* 6:248–61.

Lewis, R. N., and R. N. McElhaney. 2009. The physicochemical properties of cardiolipin bilayers and cardiolipin-containing lipid membranes. *Biochim. Biophys. Acta* 1788:2069–79.

Liebman, P. A., and G. Entine. 1974. Lateral diffusion of visual pigment in photoreceptor disk membranes. *Science* 185:457–59.

Litman, B. J., S.-L. Niu, A. Polozova, et al. 2001. The role of docosahexaenoic acid containing phospholipids in modulating G protein-coupled signaling pathways: visual transduction. *J. Mol. Neurosci.* 16:237–242; discussion 279–84.

Poo, M., and R. A. Cone. 1974. Lateral diffusion of rhodopsin in the photoreceptor membrane. *Nature* 247:438–41.

Porter, R. K., and M. D. Brand. 1993. Body mass dependence of H+ leak in mitochondria and its relevance to metabolic rate. *Nature* 362:628–30.

Soubias, O., and K. Gawrisch. 2007. Docosahexaenoyl chains isomerize on the sub-nanosecond time scale. *J. Am. Chem. Soc.* 129:6678–79.

Stillwell, W., L. J. Jenski, F. T. Crump, and W. Ehringer. 1997. Effect of docosahexaenoic acid on mouse mitochondrial membrane properties. *Lipids* 32:497–506.

Stillwell, W., and S. K. Wassall. 2003. Docosahexaenoic acid: membrane properties of a unique fatty acid. *Chem. Phys. Lipids* 126:1–27.

Valentine, D. L. 2007. Adaptations to energy stress dictate the ecology and evolution of the Archaea. *Nat. Rev. Microbiol.* 5:316–23.

Valentine, R. C., and D. L. Valentine. 2009. *Omega-3 fatty acids and the DHA principle.* Boca Raton, FL: Taylor and Francis Group.

11 Mitochondrial Membranes as a Source of Reactive Oxygen Species (ROS)

Three sources contributing to the pool of ROS in animal cells are discussed in this chapter—ROS production by macrophages, respiration, and lipid peroxidation of mitochondrial polyunsaturated membranes. There is still some debate about the major pathways of production of ROS in the cell. This state of affairs is due in part to the multiplicity of routes of ROS production, complex chemistry of ROS, great variability and modulation of rates of formation of ROS by a given pathway, technical difficulty in quantifying ROS production in living organisms, as well as multiple and often redundant mechanisms of ROS detoxification. The case of ROS production by macrophages was chosen because it is possible to definitively identify these disease-fighting cells as a major beneficial source of ROS for killing pathogens in humans. However, chronic or excessive ROS production by macrophages can cause peripheral damage to cells, triggering disease.

11.1 THE BENEFICIAL ROLE OF ROS PRODUCTION BY PHAGOCYTIC CELLS IS TO KILL PATHOGENS BY INFLICTING CATASTROPHIC OXIDATIVE DAMAGE, BUT THERE ARE SIDE EFFECTS

Data from studies of ROS production by phagocytic cells provide many lessons on the production and bioactivity of ROS. A first lesson is that bursts of ROS production by these disease-fighting cells are not always beneficial since some peripheral damage usually occurs. A second lesson is that cells that produce chemicals as harsh as ROS must protect their own membranes and mitochondria against these toxic molecules. The third lesson is that phagocytic cells can trigger oxidative bursts of ROS in the absence of an external pathogenic invader. This defines the process of sterile inflammation, in which conditions in the cell in the absence of a pathogen trigger an oxidative burst of ROS, a process that can inflict peripheral damage to healthy cells (Weitzman et al., 1985).

The membrane-embedded form of NADPH oxidase in macrophages (Figure 11.1) is responsible for the classic oxidative burst forming two superoxide radicals per molecule of NADPH consumed, as shown below:

$$NADPH + 2O_2 \xrightarrow[\text{oxidase}]{\text{NADPH}} NADP^+ + 2O_2 \cdot{-}$$

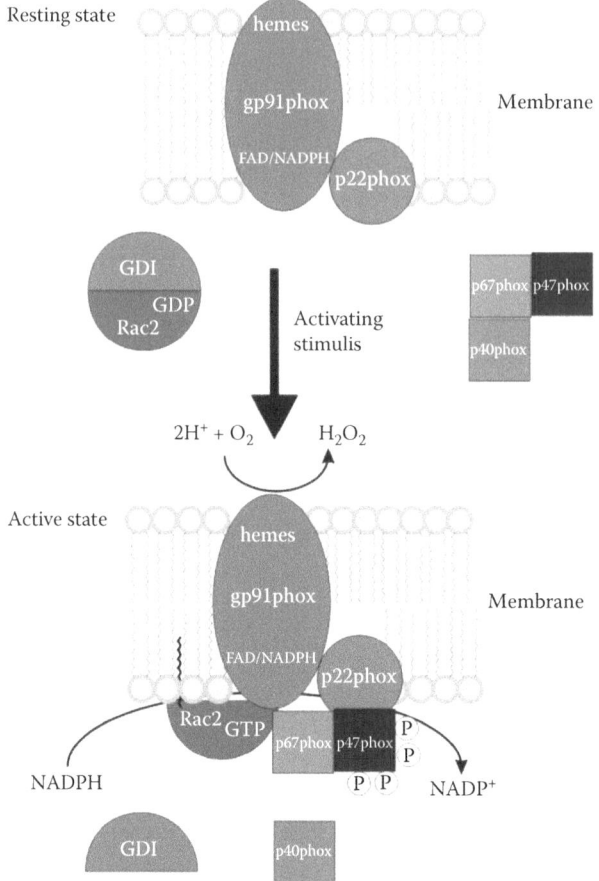

FIGURE 11.1 Schematic illustration of the activation of NADPH oxidase in phagocytic cells. NADPH oxidase activity is modulated by a complex regulatory system that involves the G-protein Rac. In resting cells a membrane embedded heterodimer of two polypeptides (p22-phox and gp91-phox), which also contains two heme groups as well as a FAD group, enables the transfer of electrons from cytosolic NADPH across the membrane to molecular oxygen without NADPH oxidase activity. Upon stimulation, a number of polypeptides (p47-phox, p67-phox, and p40-phox) translocate to the inner face of the plasma membrane to form a fully active enzyme complex capable of generating an oxidative burst of ROS. (From Held, *An Introduction to Reactive Oxygen Species: Measurement of ROS in Cells*, BioTek Instruments, 2010. Copyright © 2010 BioTek Instruments, Inc., Winooski, VT.)

NADPH oxidase is the key enzyme in the oxidative burst and is located as a large membrane-embedded enzyme complex formed by six different subunits. NADPH, the electron donor, is considered a more potent source of high-energy electrons for generation of superoxide radicals than NADH because of maintenance by the macrophage of a relatively high ratio of reduced to oxidized forms inside the cell. Two electrons from NADPH are donated to NADPH oxidase, likely in hydride transfer to flavin adenine dinucleotide (FAD), where single electrons are routed via a low potential cytochrome to O_2, forming two superoxide ($O_2^{\cdot-}$) radicals (Segal et al., 1992). Hydrogen peroxide (H_2O_2), not shown, is a major product of an oxidative burst and can be formed chemically or via superoxide dismutase from superoxide. Proof that the purpose of an oxidative burst is to kill pathogens comes from studies of the genetic chronic granulomatous disease, where in this disease the lack of or deficient levels of NADPH oxidase result in decreased ROS production and poor clearance of many bacterial and fungal pathogens.

Medical microbiologists have used another approach to study how pathogens protect themselves against the production of ROS by neutrophils, data that further define the importance of ROS in cellular metabolism. For example, the ubiquitous ulcer-forming bacterium *Helicobacter pylori* induces a strong inflammatory response in the lining of the stomach, which attracts phagocytes. This pathogen has evolved several mechanisms to prevent lipid peroxidation of its membranes. These include the following:

- Catalase to detoxify H_2O_2.
- Fatty acid hydroperoxidase, which degrades membrane fatty acid peroxides as they are formed.
- Superoxide dismutase, preventing buildup of superoxide radical and generating the less toxic form of ROS, H_2O_2.
- A ferritin-like protein that sequesters free iron, which is known as a catalyst of membrane peroxidation.
- A novel NADPH-quinone reductase that rapidly shuttles high-energy electrons to quinones (Wang and Maier, 2004), decreasing H_2O_2 production from the electron transport chain by bypassing complex 1, likely the major source of ROS within the pathogen.
- Cyclopropane coverage of double bonds of fatty acids as a mechanism against peroxidation.
- Production of membrane fatty acids with a maximum of a single double bond per chain as a mechanism to minimize peroxidation.
- The pathogen is able to grow efficiently at about one-third ambient levels of O_2; note that rates of membrane peroxidation respond linearly to O_2 levels (see Chapter 12).

Targeted genetic knockout of catalase results in nonvirulent strains of *Helicobacter hepaticus* (Hong et al., 2007), and to compensate, this mutant up-regulates synthesis of fatty acid hydroperoxidase. Catalase is strategically targeted to both the cytoplasm and periplasm of this gram-negative pathogen, and thus is able to intercept external or internally generated H_2O_2. Catalase-minus mutants of *Helicobacter* sp. display elevated rates of membrane peroxidation, increased DNA damage, and higher levels of mutations. The biochemistry of *Helicobacter* sp. establishing several of the points listed above is reviewed by Wang and colleagues (2004).

Recently, Charoenlap and colleagues (2012) showed that a knockout of *Helicobacter cinaedi*'s antioxidant protein, alkyl hydroperoxide reductase, renders the mutant more sensitive to oxidative killing by macrophages. These data are consistent with the membrane of the pathogen as a major target of oxidative damage inflicted by macrophages; detoxifying the dysfunctional membrane building blocks once they are formed ensures the survival of the pathogen.

In their normal pathogen-fighting function, neutrophils produce a battery of ROS molecules, including superoxide radical, H_2O_2, hydroxyl radical, and hypochlorous acid (HOCl). HOCl is formed by myeloperoxidase and is a major killer of pathogens. The mode of action of HOCl may involve energy uncoupling of target pathogens caused by peroxidation of membranes (Güngör et al., 2010).

There is increasing attention to the linkage among chronic infections, inflammation, and cancer. Since both the pathogen and host cells are exposed to ROS for long periods during chronic infections, their chromosomal DNA might be a target of ROS-mediated damage (i.e., mutations) causing cancer. Thus, pathogenic bacteria as reporters of oxidatively damaged DNA might shed light on the role and importance of ROS acting directly as a mutagen (Touati, 2010; Ferguson, 2010). This is an important area for future research linking ROS, cancer, and aging.

11.2   MITOCHONDRIA AS A MAJOR SOURCE OF ROS

The hypothesis that O_2 pirates electrons from the respiratory chain of mitochondria as a major cellular source of reactive oxygen species is a cornerstone of the free radical theory of aging. Estimates for mitochondrial ROS production originally ran as high as 1 to 2 percent of total O_2 used during respiration being consumed to produce ROS, but this estimate is likely orders of magnitude too high (see Murphy, 2009). There is little doubt ROS are produced as a by-product of the electron transport chain in isolated mitochondria, but details of in vivo production remain controversial, including the major sites, in vivo rates of production, regulation, and even the biological consequences of respiratory-derived ROS (Murphy, 2009; Cochemé and Murphy, 2008; Brand, 2010; Murphy et al., 2011). Recent data show that molecular oxygen pirates high-energy electrons at the flavin mononucleotide (FMN) site of complex 1 generating a superoxide radical followed by hydrogen peroxide (Figure 11.2a) (Pryde and Hirst, 2011). NADH:ubiquinone oxidoreductase (complex 1) is considered to be a major source of reactive oxygen species in mitochondria, and is thus a major contributor to cellular oxidative stress. In test tube studies using the isolated complex 1, the reduced flavin is known to react with molecular oxygen to form predominantly superoxide. High-energy electrons donated from NADH enter the electron transport arm of complex 1 at the FMN site. The initial electron transfer step involves two electrons donated to FMN in the form of a hydride transfer. Note that the molecular structure of NADH significantly shields its own cargo of two electrons from spontaneous transfer to O_2. FMN splits the initial pair of electrons donated by NADH into single high-energy electrons in preparation for entry into the electron tunnel of complex 1, an elaborate channel composed of seven successive iron-sulfur clusters, each accepting and handling a single electron at a time.

O_2 snatches single electrons at the FMN site, forming predominantly superoxide; less than 10 percent of the oxygen is reduced directly to H_2O_2. Superoxide, once formed, is dismutated rapidly to produce H_2O_2. In practice, virtually all of the superoxide produced by complex 1 is converted to H_2O_2. Thus, the rate of formation of

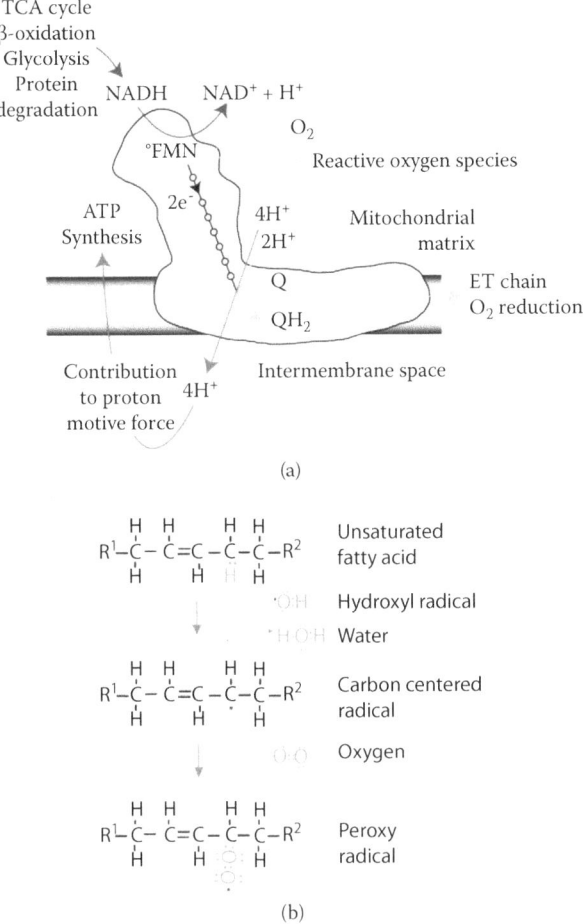

FIGURE 11.2 Complex 1 as a major source of reactive oxygen species. (a) Electrons from NADH are normally transferred to FMN and then pass through a series of iron-sulfur clusters to ubiquinone, driving the electron transport chain of mitochondria. However, reduced FMN of complex 1 also leaks high-energy electrons directly to O_2, forming reactive oxygen species, including superoxide and H_2O_2. There are increasing data suggesting that complex 1 is the major source of ROS in the cell. (From Hirst, *Biochem. J.* 425:327–39, 2010. Copyright © The Biochemical Society. Reproduced with permission.) (b) Highly polyunsaturated mitochondrial membranes in the presence of oxygen are targets of peroxidation and can generate a chain reaction. See text for details. (From Held, *An Introduction to Reactive Oxygen Species: Measurement of ROS in Cells*, BioTek Instruments, 2010. Copyright © 2010 BioTek Instruments, Inc., Winooski, VT.)

both superoxide and H_2O_2 by complex 1 can be accurately measured by quantifying the rate of H_2O_2 production. It is important to note that the rate of superoxide production by the flavin in complex 1 is set by the existing equilibrium with NADH and NAD^+. This ratio between the reduced and oxidized forms of NAD^+ is a measure of the reducing power being donated from NADH. Standard midpoint potentials listed in textbooks show a redox potential of -0.32 V for a 50–50 mixture of NADH to NAD^+. However, at a theoretical ratio of 100 NADH to 1 NAD^+ (an unlikely condition in mitochondria), the reducing power of this mixture is significantly increased, being roughly equivalent to the reducing power of hydrogen gas (-0.42 V).

As mentioned above, in the case of macrophages, the ratio of NADPH to $NADP^+$ in the cell is usually maintained higher than that of NADH to NAD^+, which means that NADPH is generally a stronger reductant than NADH, even though their midpoint potentials are similar. This helps explain why NADPH, rather than NADH, has evolved as the electron donor of choice to drive beneficial oxidative bursts of superoxide by phagocytic cells. However, mitochondria produce large amounts of NADH from energy metabolism, so there is never a shortage, although ratios of NADH to NAD^+ can vary. The ratio of NADH to NAD determines the population of complex 1 molecules that have a fully reduced flavin. The electron leak from reduced FMN to O_2, forming superoxide, is closed when the NADH site on the enzyme is occupied by NAD. Thus, complex 1 is largely able to protect electron leakage to O_2, except in one state, in which the reduced FMN site is no longer protected by a bound pyridine nucleotide. Even in its most vulnerable state, ROS production by complex 1 is relatively slow compared to rates of electron flow through the respiratory chain, but is apparently very important, being perhaps the major source of reactive oxygen species in the cell.

The flavin site mechanism developed by Pryde and Hirst (2011) explains how any event that raises or lowers the mitochondrial NAD^+ pool may govern ROS production with important consequences for longevity (see Chapter 16). These researchers point out that their unified mechanism of regulation of ROS production by complex 1 helps explain why expression of the yeast alternative (non-energy-transducing) NADH dehydrogenase expressed in a recombinant of *Drosophila* confers increased life span. They propose that the operation of a foreign and energy-wasteful NADH dehydrogenase drains the NADH pool by bypassing complex 1, raising the NAD+/NADH ratio while lowering ROS production. The net effect is increased longevity of *Drosophila* (Sanz et al., 2010).

The flavin site mechanism may also help explain the data of Hagopian and colleagues (2010), who found a dramatic decrease in ROS production by complex 1 in fat-1 mice overproducing DHA/EPA and elevating highly unsaturated fatty acid (HUFA) enrichment of their mitochondrial membranes. We suggest a possible mechanism in which an increase in membrane levels of DHA/EPA turbocharges the electron transport chain of mitochondria (see Chapter 9), raising the NAD^+/NADH ratio and thereby lowering electron leaks to form ROS. In essence, the data of Hagopian and coworkers (2010) suggest that mice have evolved a sophisticated and dynamic DHA/EPA-mediated regulatory system that down-regulates ROS production at the level of complex 1. Specifically, DHA/EPA, each well known as a target of peroxidation, acts as a negative feedback loop by lowering rates of ROS production by complex 1.

This is a win-win-lose situation from a bioenergetic perspective. Enhanced energy caused by DHA/EPA is a win, as is the decreased threat of oxidative stress. But these benefits are balanced against risks associated with the sensitivity of DHA/EPA to peroxidation, followed by membrane dysfunctions. These data also reinforce the concept that it is very important for the health of the cell to protect its unsaturated mitochondrial membranes from oxidative damage, as reinforced by the evolution of multiple mechanisms to protect membranes (Section IV). It is interesting to speculate that the dual roles of DHA/EPA, acting as both turbochargers of electron transport and down-regulators of ROS production, provide a selective advantage for a mouse, which is both short-lived and dependent on hyperactive respiration. Recall that DHA and other HUFAs, while present in mitochondria of mice, are largely absent from mitochondria of humans and other long-lived mammals. The risks associated with DHA peroxidation leading to membrane dysfunction might have less impact on a mouse, because of its short life span, than on a long-lived human. Turning this argument around, DHA-enriched mitochondria might have a marked selective disadvantage in the evolution of longevity of humans (see Chapter 14).

11.3 MITOCHONDRIAL POLYUNSATURATED MEMBRANES AS A MAJOR SOURCE OF ROS

Mitochondrial membranes are proposed to be an important source of ROS in the cell (Figure 11.2b). A linkage between ROS from complex 1 and facilitation of membrane peroxidation as a source of ROS is discussed next. We first explore the topology of ROS production measured from the FMN site (high-energy electron site) of complex 1 to the nearest membrane double bonds acting as potential targets of peroxidation (Chapter 12). The following distance parameters help explain a possible linkage between H_2O_2 produced by complex 1 and membrane peroxidation of mitochondrial membranes, leading to a vicious cycle of oxidative membrane damage mediated by ROS:

- The FMN site of complex 1 generating ROS is located on the electron transport arm of complex 1 at about 90 Å from the surface of the inner leaflet of the polyunsaturated mitochondrial membrane.
- H_2O_2 produced by complex 1 is both water soluble and lipid soluble, readily penetrating into the unsaturated region of the membrane.
- The nearest double bonds of the membrane fatty acid chains acting as targets of ROS are estimated to be located at a distance of about 100 Å from the site of origin of ROS on complex 1.
- Fenton chemistry involving H_2O_2/Fe^{+2} (see Valentine and Valentine, 2013) is proposed to generate powerful oxidizing conditions in or on the membrane, promoting lipid peroxidation via a chain reaction generating more ROS.
- Mitochondrial membrane peroxidation is thus proposed to contribute significantly to the total cellular pool of ROS.

More than 110 years ago H. J. H. Fenton discovered that a mixture of H_2O_2/Fe^{+2} generates a powerful oxidizing agent capable of degrading organic acids (Prousek, 2007). Fenton reagent features the potent oxidizing effect of oxidative free radicals such as hydroxyl radical (HO•). The decomposition of H_2O_2 catalyzed with ferrous or ferric iron is a chain reaction in which Fe^{+2} is regenerated. Trace levels of free iron in the cell are sufficient to catalyze the oxidative decomposition of unsaturated membrane fatty acids.

Membrane peroxidation is a chain reaction resulting in the formation of fatty acid peroxides (Figure 11.2b), the organic chemical equivalent of H_2O_2. Recall that peroxidation of polyunsaturated membranes, typical of human cells and mitochondria, is dependent on molecular oxygen. But O_2 does not directly attack the double bonds of fatty acid chains, and instead depends on an initiation step to start the peroxidation reaction. H_2O_2 is lipid soluble and can enter the membrane of mitochondria, say, from a nearby production site on complex 1. H_2O_2 is a potential initiator of peroxidation, but first must be converted to the powerful oxidant hydroxyl radical (HO•). HO• can readily extract an electron from the double bond of a fatty acid chain (RH), generating first a fatty acid free radical (R•), which then reacts with oxygen as follows:

$$RH + HO• \rightarrow R• + H_2O$$

$$R• + O_2 \rightarrow ROOH$$

ROOH is shorthand for a fatty acid peroxide with chemical properties similar to H_2O_2. Thus, within the relatively short distance of 100 Å or so from a potential major source of ROS, the following reactions might occur, adding to the pool of ROS as follows:

$$O_2 \rightarrow O_2• \rightarrow H_2O_2 \rightarrow HO• \rightarrow R• \rightarrow ROOH$$

A chain reaction occurs when peroxyl radical (ROO•) is generated from ROOH, as follows:

$$ROOH \rightarrow ROO•$$

ROO• (peroxyl radical) gives rise to the name membrane fatty acid peroxidation.

It is difficult to distinguish which source of ROS—electron transport or membrane—is the greatest threat to the cell. However, we suggest that polyunsaturated membranes of mitochondria are of great importance as sources of ROS because membrane peroxidation, damaging both the membrane and its embedded proteins, can potentially decrease energy production while generating a chain reaction. Membranes as targets of peroxidation are discussed in more detail in Chapter 12.

11.4   SUMMARY

The original free radical or oxidative hypothesis of aging states that ROS generated by mitochondria as well as other sources can directly mutagenize mtDNA and damage other vital cellular macromolecules, causing aging. Recent data suggest that the heart of this hypothesis dealing with the ROS $\rightarrow$ mtDNA mutation step requires revision (Chapters 6, 8, and 9). Among the three essential biopolymers most often associated with oxidative damage—DNA, proteins, membranes—membranes are gaining increasing attention. It has been known for decades that unsaturated cellular membranes not only are targets of oxidative damage causing membrane dysfunction, but also act as important sources or contributors to the overall cellular ROS pool. The current picture is that mitochondria produce most of the ROS in the cell, whereas membranes, while contributing to the ROS pool, are major targets of ROS-mediated damage, as discussed in Chapter 12.

REFERENCES

Brand, M. D. 2010. The sites and topology of mitochondrial superoxide production. *Exp. Gerontol.* 45:466–72.

Charoenlap, N., Z. Shen, M. E. McBee, et al. 2012. Alkyl hydroperoxide reductase is required for *Helicobacter cinaedi* intestinal colonization and survival under oxidative stress in BALB/c and BALB/c interleukin-10–/– mice. *Infect. Immun.* 80:921–28.

Cochemé, H. M., and M. P. Murphy. 2008. Complex I is the major site of mitochondrial superoxide production by paraquat. *J. Biol. Chem.* 283:1786–98.

Ferguson, L. R. 2010. Chronic inflammation and mutagenesis. *Mutat. Res.* 690:3–11.

Güngör, N., A. M. Knaapen, A. Munnia, et al. 2010. Genotoxic effects of neutrophils and hypochlorous acid. *Mutagenesis* 25:149–54.

Hagopian, K., K. L. Weber, D. T. Hwee, et al. 2010. Complex I-associated hydrogen peroxide production is decreased and electron transport chain enzyme activities are altered in n-3 enriched fat-1 mice. *PLoS One* 5(9):e12696.

Hirst, J. 2010. Towards the molecular mechanism of respiratory complex I. *Biochem. J.* 425:327–39.

Hong, Y., G. Wang, and R. J. Maier. 2007. A *Helicobacter hepaticus* catalase mutant is hypersensitive to oxidative stress and suffers increased DNA damage. *J. Med. Microbiol.* 56:557–62.

Murphy, M. P. 2009. How mitochondria produce reactive oxygen species. *Biochem. J.* 417:1–13.

Murphy, M. P., A. Holmgren, N. G. Larsson, et al. 2011. Unraveling the biological roles of reactive oxygen species. *Cell Metab.* 13:361–36.

Prousek, J. 2007. Fenton chemistry in biology and medicine. *Pure Appl. Chem.* 79:2325–38.

Pryde, K. R., and J. Hirst. 2011. Superoxide is produced by the reduced flavin in mitochondrial complex I: a single, unified mechanism that applies during both forward and reverse electron transfer. *J. Biol. Chem.* 286:18056–65.

Sanz, A., M. Soikkeli, M. Portero-Otín, et al. 2010. Expression of the yeast NADH dehydrogenase Ndi1 in *Drosophila* confers increased lifespan independently of dietary restriction. *Proc. Natl. Acad. Sci. USA* 107:9105–10.

Segal, A. W., I. West, F. Wientjes, et al. 1992. Cytochrome b-245 is a flavocytochrome containing FAD and the NADPH-binding site of the microbicidal oxidase of phagocytes. *Biochem. J.* 284:781–88.

Touati, E. 2010. When bacteria become mutagenic and carcinogenic: lessons from *H. pylori*. *Mutat. Res.* 703:66–70.

Valentine, R. C., and D. L. Valentine. 2013. *Neurons and the DHA principle*. Boca Raton, FL: Taylor and Francis Group.

Wang, G., R. C. Conover, S. Benoit, et al. 2004. Role of a bacterial organic hydroperoxide detoxification system in preventing catalase inactivation. *J. Biol. Chem.* 279:51908–14.

Wang, G., and R. J. Maier. 2004. An NADPH quinone reductase of *Helicobacter pylori* plays an important role in oxidative stress resistance and host colonization. *Infect. Immun.* 72:1391–96.

Weitzman, S. A., A. B. Weitberg, E. P. Clark, et al. 1985. Phagocytes as carcinogens: malignant transformation produced by human neutrophils. *Science* 227:1231–33.

12 Mitochondrial Membranes as Major Targets of Oxidation

Potential targets of oxidative damage in the cell have recently been reviewed (Figure 12.1) (Murphy et al., 2011; Murphy, 2009). For many years it was widely accepted that a major pathway for amplifying reactive oxygen species (ROS)-mediated damage causing aging involves a mechanism in which ROS act directly to mutagenize mtDNA. ROS, such as generated by mitochondria, were envisioned to directly attack the bases of mtDNA-generating mutations, thus causing aging. Recent data suggest that errors of replication of mtDNA, not ROS, generate mutations that cause aging (see Chapter 9). These data open Pandora's box concerning what fundamental roles, if any, oxidative stress mediated by ROS plays in aging. The purpose of this chapter is to make a single point—membranes deserve more attention as mediators of aging. Case histories are chosen to document that membranes are clearly major targets of oxidative damage. We suggest that mitochondrial membranes of humans belong to a class of specialized membranes sensitive to per-oxidation; the result is a membrane dysfunction so powerful as to trigger apoptosis.

12.1 RHODOPSIN MEMBRANE DISKS ARE HIGHLY ENRICHED WITH DHA AND AGE RAPIDLY

It is well known that specialized membrane disks essential for sensing light in rod cells of the eye turn over rapidly in animals (Young, 1967). The light-sensing protein rhodopsin is the predominant protein embedded in these membranes, and this gave rise to their name, rhodopsin membrane disks. Rhodopsin membrane disks have served as a model system in the field of membrane peroxidation and have provided important data relevant to understanding the oxidative stability/instability of highly polyunsaturated mitochondrial membranes. Some selected highlights from research on rhodopsin membrane disks as targets of lipid peroxidation are as follows:

- Docosahexaenoic acid (DHA) levels as high as 53 percent of total fatty acids have been reported in rhodopsin membrane disks in certain rodents.
- Photooxidation in addition to peroxidation greatly accelerates membrane damage.
- The rate of senescence of rhodopsin disks in frogs increases markedly and linearly with increasing temperature, and vice versa.

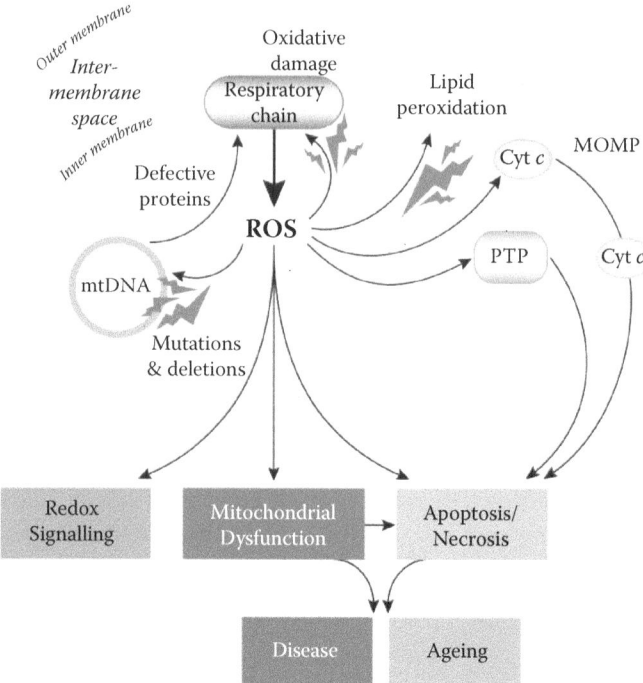

FIGURE 12.1   See color insert. Overview of major cellular targets of oxidative damage. This diagram shows that electron leaks from the respiratory chain of mitochondria form ROS, which damage membrane lipids, generate defective proteins, and act as signaling molecules for opening permeability transition pore (PTP) and releasing membrane-bound cytochrome C (cytC). (From Murphy, *Biochem. J.* 417:1–13, 2009. Copyright © 2009, The Biochemical Society. Reproduced with permission.)

- Rhodopsin membrane disks in mammals contain high levels of di-DHA molecular species of phospholipids. With 12 double bonds per molecule, these membrane building blocks are among the most sensitive to oxidation in nature.
- Roughly half of DHA chains in a senescing rhodopsin membrane disk are damaged by oxidation; the remainder is recycled.
- Truncated DHA phospholipids as by-products of lipid peroxidation have been identified and synthesized.
- Rhodopsin protein and other membrane proteins are readily inactivated in aging disks by reactive aldehydes generated as by-products of membrane fatty acid peroxidation.
- Excessive DHA incorporated into rhodopsin disks of transgenic mice over-expressing DHA genes is harmful to vision during exposure to a regiment of bright light.

- Oxidation products of DHA such as found in damaged rhodopsin disks react with DNA and score positive in the Ames mutagen test (Chapter 3).
- Research using rhodopsin disks has played an important role in development of the lipid whisker model of oxidatively damaged membrane surfaces (Chapter 13).

Rhodopsin membrane disks present in the outer segment in rod cells of the eye are targets of both photooxidation and peroxidation, the former during waking periods (i.e., when the eye is open) and the latter in the dark (reviewed by Giusto et al., 2000). Due to their high DHA content, it is predictable that these specialized bilayers (located in a stack of about 1000 disks in the outer segment of rod cells) are unstable and require continuous renewal. The life span of a newly minted membrane entering the bottom of the stack and exiting at the apex is about 10 days (Figure 12.2). That is, roughly 100 disks, or 10 percent of total disks, are removed as spent or senescing

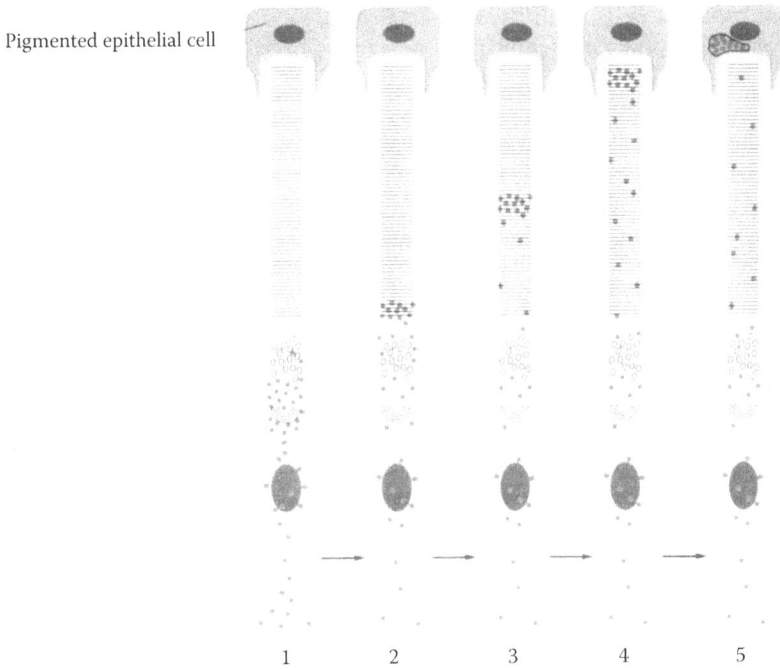

Pigmented epithelial cell

1   2   3   4   5

FIGURE 12.2   Ten-day cycle of renewal of rhodopsin disks in rod cells of the eye. In 1967 Young discovered the rapid turnover of rhodopsin disk membranes using a radioactive pulse-labeling method. Newly minted disks enter the bottom of the stack, and aged disks reaching the top of the stack are recycled. Renewal of the entire stack of rhodopsin disks takes about ten days. These data, revolutionary at the time, can now be explained by the presence of about 50 percent DHA as total fatty acids in mammalian membrane disks, along with the powerful effects of photooxidation and peroxidation causing degradation of these highly unsaturated membranes. (From Alberts et al., *Molecular Biology of the Cell*, 3rd ed., 1994. Copyright © 1994. Garland Science-Books, NY. Republished with permission of Garland Science-Books; permission conveyed through Copyright Clearance Center, Inc.)

disks from the top of the stack daily, being replaced with an equal number of newly minted disks on the bottom.

In this classic experiment summarized in Figure 12.2, R. W. Young (1967) was able to "tag" newly minted membrane disks entering the bottom of the stack of rhodopsin disk membranes with a pulse of radioactive amino acid (shown as dots in the diagram), which is incorporated primarily into rhodopsin protein. In this experiment an intense band of radioactivity is first seen in newly minted disks entering the bottom of the stack. In this case, it is not DHA that is labeled but newly synthesized rhodopsin protein, which is inserted into newly synthesized disks and acts as a marker to follow the upward progression of disks over time. During a period lasting about ten days, the band of radioactivity moves from the bottom to the top of the stack, where spent or "old" disks are removed. Loss of radioactive rhodopsin shows that these membranes are being continually recycled at a remarkable pace. Note that data from this pulse-labeling experiment simultaneously measures the functional life span of both rhodopsin and the DHA-enriched membranes that embed this protein. Young, in his classic experiments on the rapid turnover of rhodopsin disk membranes, showed that increasing the temperature from 14°C to 34°C accelerated decay of rhodopsin disks by 2.5-fold. This is a striking result given that his experiments involved living animals held at different temperatures. In replotting Young's original data (Young, 1967), it appears that rates of membrane senescence (oxidation) increase roughly linearly with temperature from 4 to 34°C, as predicted by chemical theory.

In humans, conditions of warm body temperatures, light, and high fluxes of O_2, combined with the high DHA composition of rod cell disks, make these rhodopsin disk membranes ideal substrates for chemical attack by oxygen. The retina has evolved several defense mechanisms that protect against oxidative damage. Conventional defenses include high levels of protective enzymes (found in photoreceptor outer segments) that intercept and destroy reactive oxygen species before they attack rhodopsin membranes. Antioxidants are a main line of defense against autoxidation in the retina (Robison et al., 1982; Valk and Hornstra, 2000). Vitamin E appears to be located strategically in rhodopsin disks and protects DHA from oxidation. Researchers have shown that incorporation of vitamin E into DHA-enriched membrane vesicles significantly reduces damage to DHA.

A variety of other antioxidants also help protect our vision. Even though the retina has evolved numerous defenses usually efficient enough in most cells to prevent oxidative damage, the presence of high levels of DHA, especially in the presence of visible light, appears capable of overwhelming these defensive systems in rod cells. Note that photooxidation is not a free radical reaction and is not terminated by antioxidants. This necessitates additional protection of rhodopsin disks in the form of active remodeling and highly sophisticated membrane renewal mechanisms. In addition, many other mechanisms have likely evolved to protect mitochondrial membranes of rod cells (see Section IV). With aging or through environmental insults, defenses in the eye against oxidative damage might begin to deteriorate, resulting in death of rod cells and serious damage to vision. For example, at high light intensities or elevated enrichment with DHA, the risks associated with DHA begin to outweigh the benefits, helping trigger a pathological state damaging vision.

Preservation of our vision depends on yet another sophisticated system. The cellular physiology of renewal of rhodopsin disks involves an elaborate process with spent disks being shed, then taken up and processed by specialized scavenging cells called pigmented epithelial cells. See Chapter 13 for roles of lipid whiskers in signaling phagocytic cells. Within pigmented epithelial cells, fragments of senescing disks become surrounded by cell membranes to form inclusions called phagosomes. The phagosomes are subsequently degraded within the pigmented epithelial cells, with undamaged DHA being salvaged and used for new disk synthesis. Phagocytosis of spent disks is accelerated as expected when the disks in the intact eye are exposed to excessive light, with prolonged intense light causing permanent damage and, in extreme cases, blindness. As mentioned above, the rate of turnover of rhodopsin disks in rod cells of frogs' eyes drops sharply as temperatures are lowered from 34°C to 4°C (Young, 1967), which is consistent with cool temperatures characteristic in their environment likely acting to strongly decrease rates of oxidative damage to DHA membranes. Overexpression of DHA synthesis in transgenic mice has the same effect as too much light, accelerating aging of disks, with collateral damage spreading and resulting in death of retinal cells.

12.2 LESSONS FROM DHA-ENRICHED TAILS OF SPERM

In humans three kinds of specialized cells have evolved membranes that are extremely sensitive to attack by ROS and require extraordinary protection—photoreceptor cells (already discussed), sperm, and neurons. The general importance of DHA as a building block for healthy sperm is well established, but with a surprise: the strategic localization or targeting of DHA to the tail membrane (Connor et al., 1997). Indeed, one of the most unusual chemical features of sperm membranes is the differential enrichment of the tail membrane versus head membrane, with 19.6 percent DHA in the tail compared to 1.1 percent in the head based on analysis of monkey sperm. These values are expected to be similar in human sperm. Arachidonic acid (20:4) is also preferentially targeted to the tail of monkey sperm—6.4 percent in the tail versus 1.6 percent in the head. The overall differences in total unsaturated fatty acids are 34.1 percent in the tail versus 12.1 percent in the head membrane. These data support the concept discussed in Chapter 14 that DHA can be targeted to specific membrane domains.

Studies on spermatogenesis show that sperm production is an imperfect process under the best of conditions, subject to numerous forms of environmental stress. In normal semen, about 20 million sperm are present per cubic centimeter, of which about 5 million are defective. The sensitivity of sperm biogenesis to environmental stresses helps explain why sperm counts may plummet following sessions in saunas, hot tubs, or after wearing tight-fitting cycling shorts, each of which causes heat buildup around the scrotum and destruction of sperm. Indeed, the external location of the scrotum is thought to have evolved as a cooling mechanism, perhaps against membrane peroxidation, and leading to higher levels of viable sperm. The fact that sperm tail membranes are highly enriched with DHA plays into this scenario, raising the likelihood that protection against membrane peroxidation might be important in male fertility. It is interesting to note that naked mole rats, which remain sexually active through old age, maintain their body temperature in the range of 30 to 32°C (Chapter 6). Yet in spite of the relatively cool temperatures of their scrotums, the ratio of dysfunctional to active sperm of naked mole rats is high, comparable to sperm counts found in DHA knockout mutants of mice. DHA deficiency in mice causes infertility. These data raise the possibility that super-males might arise in colonies of naked mole rats and be the main breeders with the queen.

Chemical studies on animal and human sperm established that DHA-enriched phospholipids are highly susceptible to oxidation, and this process is associated with progressive and irreversible loss of structural integrity, motility, viability, and metabolic activity of the sperm cells (Alvarez and Storey, 1995; Aitken and Clarkson, 1987). Indeed, exogenously applied lipid peroxides, the first product of lipid peroxidation, are powerfully spermicidal against washed human sperm treated with as little as 30 nM of lipid peroxide. Treated sperm become irreversibly immobile within a few minutes. Lipid peroxides are major intermediates in chemical oxidation of DHA and polyunsaturated oils and membranes. Fatty acid peroxides are not free radicals but are readily converted to lipid peroxyl radical in the presence of trace amounts of iron. Lipid peroxyl radical is a free radical and can extract an electron from a double bond of DHA in sperm tails, yielding lipid peroxide and a lipid free radical (see Chapter 11). Thus, the chemical steps of oxidation of DHA in sperm membranes can be divided into three stages, as follows:

- Initiation by Fe^{3+} or another free radical generates a fatty acid free radical.
- Propagation of lipid autoxidation in the dark occurs as a chain reaction.
- Termination occurs when available substrates are depleted or an antioxidant such as vitamin E quenches the chain reaction.

As introduced in Chapter 11, O_2 does not directly attack double bonds of fatty acid chains in sperm tails, but reacts rapidly once initiation occurs. Hydrogen peroxide (H_2O_2) damages sperm tail membranes because in the presence of Fe^{3+} the simplest form of a peroxyl radical ($H_2O_2 \rightarrow HOO\bullet$) is formed and initiates a chain reaction. This reaction shows why fatty acid peroxides are dangerous to sperm tail membranes. In other words, once fatty acid peroxides are added to a sample of washed sperm, their presence results in an oxidative chain reaction, and sperm viability is rapidly destroyed.

Classic chemical oxidation protocols using iron or ascorbate catalysis show that DHA enriched in the sperm's tail is preferentially targeted and rapidly lost from the membrane. In contrast, 16:0 remains at relatively constant levels (Alvarez and Storey, 1995), as expected based on its stability against peroxidation. Also, $18:1_{n9}$, $18:2_{n6}$, and cardiolipin, the latter diagnostic of sperm mitochondrial phospholipids present in mitochondria located in the midpiece region of sperm, are relatively stable toward this harsh lipid oxidation protocol. However, $20:4_{n6}$ is present at levels of approximately 10 percent of DHA and is subject to oxidation, though at somewhat slower rates than DHA.

The targeting of DHA to tail membranes keeps this potentially damaging source of ROS at a safe distance from the stored germ-line DNA. DHA also seems to be excluded or targeted away from membranes of sperm mitochondria. Thus, separation by distance is suggested to prevent oxidative damage to germ-line DNA caused by the accumulation of toxic ROS radicals (Fraga et al., 1991), which are readily derived from DHA-enriched sperm membranes.

Avoidance of O_2 has been shown to be a simple but effective mechanism to protect DHA-enriched cells against oxidative stress (Valentine and Valentine, 2009; also see Chapter 15). The O_2 avoidance strategy maximizes benefits of DHA and minimizes oxidative risk. The lack of most conventional ROS defenses in sperm cells is consistent with little need for such antioxidation mechanisms in the low-O_2 world found in the female reproductive tract. Note that the entire distance traveled by sperm in the reproductive tract is not devoid of O_2 since the egg cloistered in the oviduct requires respiration for survival. A review of the benefits and risks of reactive oxygen species in human sperm cell biology has recently been published and provides a comprehensive overview of this field (Koppers et al., 2008).

As discussed above, sperm tails have proven to be important research tools for studies of membrane peroxidation (see Storey, 2008, for a lucid historical account of sperm metabolism and lipid peroxidation of sperm membranes). The following conclusions from studies of peroxidation of membranes of sperm likely apply to membranes in general, especially polyunsaturated fatty acid (PUFA)-enriched membranes of mitochondria:

- Peroxidation of highly unsaturated membranes of sperm proceeds by established chemical rules.
- Rates of lipid peroxidation increase in proportion to the number of double bonds in the fatty acid chain, and vice versa (i.e., 22:6 > 20:4 > 18:2 >> 18:1).
- Lipid peroxidation rate is proportional to O_2 levels from 0 to 95 percent.
- Viability and motility of sperm decrease in proportion to oxidative membrane damage.
- Oxidative damage to sperm causes loss of membrane permeability barriers to ions, sugars, and other metabolites.
- A scrotal temperature in humans of 30 to 33°C compared to body temperatures of 37°C is estimated to lower sperm membrane peroxidation by as much as six-fold.

12.3 MITOCHONDRIAL MEMBRANES AND THEIR PROTEINS AS TARGETS OF OXIDATIVE DAMAGE

Figure 12.3 illustrates how electrons leaking from the mitochondrial electron transport chain react with O_2 generating superoxide radical, which is quickly converted to H_2O_2 (see Chapter 11). H_2O_2 is soluble in the membrane, where it can be converted by Fenton chemistry to hydroxyl radical. Hydroxyl radical can abstract an electron from a PUFA molecule generating the PUFA free radical L•. In the presence of O_2 fatty acid hydroperoxide (LOOH) is formed, which in turn is converted to protein-reactive aldehydes such as 4-HNE. Detoxification of 4-HNE occurs by conjugation with glutathione. 4-HNE acts as a protein poison forming adducts with the side

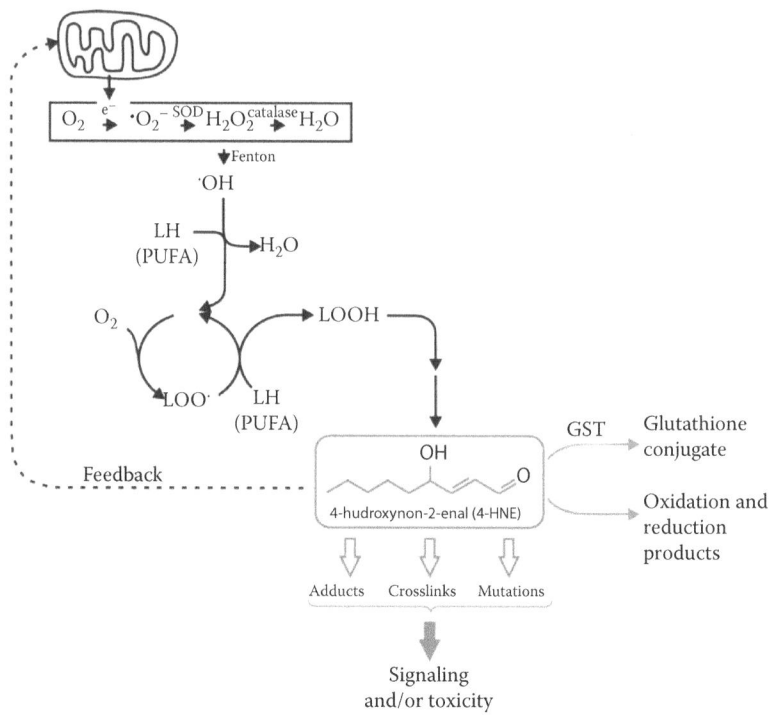

FIGURE 12.3 Proteins as targets of membrane oxidation. Reactive aldehydes such as 4-HNE capable of inactivating enzymes are products of oxidative cleavage of highly unsaturated membranes and, once formed, can directly attack proteins. Protein 4-HNE adducts can plug up essential protein quality control enzymes and destroy enzymatic activity. (From Zimniak, *Ageing Res. Rev.* 7:281–300, 2008. Copyright © 2008. Reprinted with permission from Elsevier.)

chains of amino acids, including cysteine, histidine, and lysine. Proteins that carry 4-HNE adducts can display altered function and resistance to degradation. Key protein degradation machinery (e.g., proteasomes) can be targets of reactive aldehydes or plugged up by derivatized proteins unable to pass through the proteasome. For a recent review of the multiple roles of 4-HNE in reacting with proteins and its implications for aging, see Zimniak (2008).

Zimniak (2008) points out that there are no known enzymes that remove adducts of reactive aldehydes from proteins, and this may account for the contribution of defective proteins to the "aging pigment" lipofuscin (Ng et al., 2008; Gray and Woulfe, 2005). Thus, the current picture linking membrane peroxidation and proteins is that peroxidation of unsaturated membrane fatty acids and denaturation of membrane proteins are interlinked chemical processes, together driving senescence and oxidatively mediated dysfunction of mitochondrial and other specialized membranes such as rhodopsin disks. Once again, rhodopsin disks have played an important role in understanding the synergism between membrane lipid peroxidation and degradation of membrane proteins. For example, rhodopsin protein, representing about

90 percent of total protein content of rhodopsin disks, is also a target of oxidative damage by mechanisms shown in Figure 12.3. Modern analytical and chemical synthesis techniques have allowed biochemists to use rhodopsin membranes as tools for developing a picture of interplay between damage to proteins and DHA (Hollyfield et al., 2010; Warburton et al., 2005; Lu et al., 2009). Some of the highlights are that the combination of photooxidation and peroxidation is a powerful force in destroying DHA chains present in almost half of the phospholipids in senescing rhodopsin disks. Detailed chemical analysis coupled with synthetic chemistry has led to the identification of essentially all of the degradation products predicted by chemical theory of lipid peroxidation, including potent alkylating fragments derived from DHA that can readily react with and damage proteins such as rhodopsin (Gutteridge and Halliwell, 1990; Sun and Salomon, 2004).

The final point deals with a novel class of human/animal mitochondria whose membranes are exposed to visible light, known to be a powerful catalyst promoting extremely high rates of fatty acid oxidation. Retinal ganglion cells have been implicated in the hereditary mitochondrial disease Leber's hereditary optic neuropathy (LHON), as discussed in Chapter 8. Because of their unique location, retinal ganglion cells and their axons are exposed to visible light at wavelengths of 400 to 760 nm. In studies of light damage of photoreceptors and retinal pigment epithelium, blue light has the shortest wavelength and causes the production of the highest levels of ROS. It is well known that visible light dramatically increases rates of oxidation of polyunsaturated fatty acids (Gunstone, 1996), with DHA-enriched membranes of rhodopsin membrane disks of photoreceptor cells being especially sensitive to photooxidative damage, as discussed above.

Mitochondria of most human cells are obviously shielded from photooxidation because of their location deep in the body. However, when the eye is open, mitochondrial membranes in unmyelinated retinal ganglion axons are directly exposed to light. Thus, visible light energy absorbed by triplet oxygen ($^3O_2$) in the presence of sensitizers found in retinal ganglion mitochondria is expected to generate highly reactive singlet oxygen ($^1O_2$). Singlet oxygen is known to dramatically increase oxidation rates of polyunsaturated chains, such as enriched in human mitochondrial membranes. This concept has been tested using retinal ganglion cells grown in culture and exposed to light (Lascaratos et al., 2007). Light exposure increased ROS levels and triggered the apoptotic cascade. It would be of interest to know if (18:2)-CL molecular species in light-exposed retinal ganglion cells are especially sensitive to oxidative damage.

Yu-Wai-Man and coworkers (2011) suggest that chronic light exposure could be the tipping point toward apoptosis in retinal ganglion cells already compromised by a genetically determined mitochondrial respiratory chain defect (LHON) (Paskowitz et al., 2006; see Chapter 8). These data suggest that highly polyunsaturated membranes of mitochondria, like specialized membranes of rhodopsin disks, become targets of photooxidative damage by visible light. Over time, defects caused by photooxidation seem capable of overwhelming defensive systems geared to protect mitochondrial membranes. While a majority of light-exposed retinal ganglion cells can be killed in patients with LOHN, a small subset of about 1 percent of neurons called melanopsin retinal ganglion cells are spared (La Morgia et al., 2010). These data raise questions regarding mechanisms operating in melanopsin ganglion cells that might protect their membranes from the same light-mediated mitochondrial dysfunctionality that kills retinal ganglion cells.

12.4   SUMMARY

Many of the fundamental chemical principles governing peroxidation of highly unsaturated membranes were developed many decades ago using specialized DHA-enriched membranes, especially rhodopsin membrane disks and sperm tail membranes, as research tools. As predicted by chemical theory, DHA in these membranes is highly sensitive to oxidation (Reis and Spickett, 2012). One of the predictions from these data is that highly polyunsaturated membranes of human mitochondria are also expected to be major targets of peroxidation.

Finally, there are increasing data showing that human mitochondrial membranes belong to a class of specialized membranes that are extremely sensitive to peroxidative damage. The highly polyunsaturated nature of the inner mitochondrial membrane helps explain why mitochondria divide so frequently. At first glance, the highly polyunsaturated human mitochondrial membranes can be estimated to be at least an order of magnitude more sensitive to peroxidation than, say, yeast mitochondrial membranes, which are enriched with monounsaturated fatty acids. However, data discussed in Chapter 20 suggest that one major molecular species of human mitochondrial membranes' phospholipid, cardiolipin, $(18:2)_4$-CL, may be forty- to fifty-fold more prone to peroxidation than monounsaturated membranes.

REFERENCES

Aitken, R. J., and J. S. Clarkson. 1987. Cellular basis of defective sperm function and its association with the genesis of reactive oxygen species by human sperm. *J. Reprod. Fertil.* 81:459–69.

Alvarez, J. G., and B. T. Storey. 1995. Differential incorporation of fatty acids into and peroxidative loss of fatty acids from the phospholipids of human spermatozoa. *Mol. Reprod. Dev.* 42:334–46.

Connor, W. E., R. G. Weleber, C. DeFrancesco, D. S. Lin, and D. P. Wolfe. 1997. Sperm abnormalities in retinitis pigmentosa. *Invest. Ophthalmol. Vis. Sci.* 38:2619–28.

Fraga, C. G., P. A. Motchnik, M. K. Shigenaga, et al. 1991. Ascorbic acid protects against endogenous oxidative DNA damage in human sperm. *Proc. Natl. Acad. Sci. USA* 88:11003–6.

Giusto, N. M., S. J. Pasquaré, G. A. Salvador, et al. 2000. Lipid metabolism in vertebrate retinal rod outer segments. *Prog. Lipid Res.* 39:315–91.

Gray, D. A., and J. Woulfe. 2005. Lipofuscin and aging: a matter of toxic waste. *Sci. Aging Knowledge Environ.* 2005(5):re1.

Gunstone, F. D. 1996. *Fatty acid and lipid chemistry.* London: Blackie Academic & Professional, p. 252.

Gutteridge, J. M., and B. Halliwell. 1990. The measurement and mechanism of lipid peroxidation in biological systems. *Trends Biochem. Sci.* 15:129–35.

Hollyfield, J. G., V. L. Perez, and R. G. Salomon. 2010. A hapten generated from an oxidation fragment of docosahexaenoic acid is sufficient to initiate age-related macular degeneration. *Mol. Neurobiol.* 41:290–98.

Koppers, A. J., G. N. DeIuliis, J. M. Finnie, et al. 2008. Significance of mitochondrial reactive oxygen species in the generation of oxidative stress in spermatozoa. *J. Clin. Endocrinol. Metab.* 93:3199–207.

La Morgia, C., F. N. Ross-Cisneros, A. A. Sadun, et al. 2010. Melanopsin retinal ganglion cells are resistant to neurodegeneration in mitochondrial optic neuropathies. *Brain* 133:2426–38.

Lascaratos, G., D. Ji, J. P. Wood, et al. 2007. Visible light affects mitochondrial function and induces neuronal death in retinal cell cultures. *Vision Res.* 47:1191–201.

Lu, L., X. Gu, L. Hong, et al. 2009. Synthesis and structural characterization of carboxyethylpyrrole-modified proteins: mediators of age-related macular degeneration. *Bioorg. Med. Chem.* 17:7548–61.

Murphy, M. P. 2009. How mitochondria produce reactive oxygen species. *Biochem. J.* 417:1–13.

Murphy, M. P., A. Holmgren, N. G. Larsson, et al. 2011. Unraveling the biological roles of reactive oxygen species. *Cell Metab.* 13:361–66.

Ng, K. P., B. Gugiu, K. Renganathan, et al. 2008. Retinal pigment epithelium lipofuscin proteomics. *Mol. Cell Proteomics* 7:1397–405.

Paskowitz, D. M., M. M. LaVail, and J. L. Duncan. 2006. Light and inherited retinal degeneration. *Br. J. Ophthalmol.* 90:1060–66.

Reis, A., and C. M. Spickett. 2012. Chemistry of phospholipid oxidation. *Biochim. Biophys. Acta* 1818:2374–87.

Robison, W. G., T. Kuwabara, and J. G. Bieri. 1982. The roles of vitamin E and unsaturated fatty acids in the visual process. *Retina* 2:263–81.

Storey, B. T. 2008. Mammalian sperm metabolism: oxygen and sugar, friend and foe. *Int. J. Dev. Biol.* 52:427–37.

Sun, M., and R. G. Salomon. 2004. Oxidative fragmentation of hydroxy octadecadienoates generates biologically active γ-hydroxyalkenals. *J. Am. Chem. Soc.* 126:5699–708.

Valentine, R. C., and D. L. Valentine. 2009. *Omega-3 fatty acids and the DHA principle.* Boca Raton, FL: Taylor and Francis Group.

Valk, E. E., and G. Hornstra. 2000. Relationship between vitamin E requirement and polyunsaturated fatty acid intake in man: a review. *Int. J. Vitam. Nutr. Res.* 70:31–42.

Warburton, S., K. Southwick, R. M. Hardman, et al. 2005. Examining the proteins of functional retinal lipofuscin using proteomic analysis as a guide for understanding its origin. *Mol. Vis.* 11:1122–34.

Young, R. W. 1967. The renewal of photoreceptor cell outer segments. *J. Cell Biol.* 33:61–72.

Yu-Wai-Man, P., P. G. Griffiths, and P. F. Chinnery. 2011. Mitochondrial optic neuropathies: disease mechanisms and therapeutic strategies. *Prog. Retin. Eye Res.* 30:81–114.

Zimniak, P. 2008. Detoxification reactions: relevance to aging. *Ageing Res. Rev.* 7:281–300.

Section IV

Many Mechanisms Have Evolved to Protect Human Mitochondrial Membranes, Enabling Longevity

Human cells have evolved a remarkable number of membrane-protective mechanisms that collectively prevent or minimize oxidative dysfunction of highly polyunsaturated mitochondrial membranes. The bioenergetic cost to prevent membrane oxidation is high, but we suggest that the reward is great because preventing membrane damage saves energy and enables longevity. A new generation of mitochondria-targeted antioxidants has been developed that act by blocking oxidation of cardiolipin, resulting in prevention of neuron death in rodent models of neurodegenerative diseases.

13 Apoptosis Caused by Oxidatively Truncated Phospholipids Can Be Reversed by Several Mechanisms, Especially Enzymatic Detoxification

Lipid chemists have known for many years that truncated or chain-shortened phospholipids are formed as by-products of peroxidation of unsaturated membrane lipids (Reis and Spickett, 2012). The purpose of this chapter is to explore the bioactivities of truncated phospholipids, highlighting their roles as signaling molecules triggering inflammation and apoptosis (McIntyre, 2012). The timely and comprehensive review by McIntyre provides a summary and an up-to-date reference list of the production of truncated phospholipids, with emphasis on their bioactivities. McIntyre (2012) points out that the levels of truncated phospholipids being generated in the human body may be much greater than currently appreciated because of the extremely rapid clearance or enzymatic hydrolysis of these molecules from the circulatory system. In his review, this author does not emphasize the role of truncated phospholipids in aging and age-related diseases. However, he develops a thought-provoking and well-documented model linking oxidatively truncated phospholipids and apoptosis mediated by mitochondria. In short, the data and model of apoptosis developed by McIntyre (2012) are very important in understanding the contributions of unsaturated membranes and oxidative stress in aging, especially since polyunsaturated membranes of mitochondria are likely the major source of oxidatively truncated phospholipids in the body.

13.1 LIPID WHISKERS AND THEIR GENERAL PROPERTIES

In 2008 Greenberg and colleagues (2008) developed an interesting model that helps explain how chain-shortened phospholipids alter membrane molecular architecture and function with implications for aging and age-dependent diseases. According to their lipid whisker model, truncated phospholipids display sufficient water-loving properties to move partially out of their normal position in the lipid portion of the membrane and into the aqueous phase. That is, oxidatively truncated

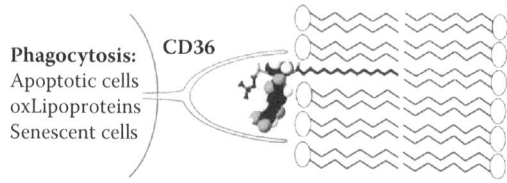

Phagocytosis:
Apoptotic cells
oxLipoproteins
Senescent cells

CD36

FIGURE 13.1 Lipid whisker model of oxidatively truncated phospholipids as signaling molecules for phagocytes and astrocytes. The physical–chemical properties of truncated phospholipids cause this normally membrane-embedded building block to emerge out of the oily bilayer extending as a molecular whisker into the aqueous phase. Truncated phospholipids act as signaling molecules for circulating phagocytes and presumably microglia of the brain. Hyperactivation of phagocytic cells by truncated phospholipids and other signaling molecules can result in sterile inflammation, causing collateral damage to neighboring cells and possibly igniting an oxidative chain reaction. The role of lipid whiskers in mitochondrial membranes is an important area for future research. (From Li et al., *Biochemistry* 46:5009–17, 2007. Copyright © 2007, American Chemical Society. Reprinted with permission.)

or chain-shortened phospholipids are amphiphilic and are proposed to display sufficient lipophilic properties to remain anchored to the membrane or, alternatively, to exit the bilayer in favor of the aqueous phase. During oxidative cleavage the former methyl end of an unsaturated chain is replaced by a hydrophilic group (Gugiu et al., 2006), which is proposed to create the dual-solubility properties of truncated phospholipids (Figure 13.1). Thus, while attached to the membrane, truncated phospholipids are proposed to stick out of the membrane's surface like tiny molecular hairs or whiskers. Some possible bioactivities of lipid whiskers are as follows:

- Lipid whiskers are in a position to signal and bind to receptors of phagocytic cells, which are triggered into a state of phagocytosis.
- Membrane repair lipases can recognize and have access to these hairs for the purpose of clearing them from the surface (Farooqui et al., 1997; Leslie, 1997).
- Lipid whiskers mark the sites of defects in the membrane surface, likely altering critical functional parameters such as membrane lateral motion and permeability, and perhaps generating water-wires, which uncouple proton gradients.
- Finally, lipid whiskers are reporters of oxidative damage to membranes and serve as biochemical markers for unhealthy or aging membranes.

Catalá (2012) has provided a picture of how the lipid whisker model and lipid raft model might be linked. Several of the major points discussed by Catalá are as follows:

- Cholesterol, sphingolipid, and fatty acids modulate the transport and diffusion of oxygen in membranes.
- Polyunsaturated phospholipids are considered to be major targets of oxidative damage.
- Vitamin E seems to be concentrated where it is most needed, in membrane regions enriched with polyunsaturated fatty acids.

- Lipid dynamics in polyunsaturated-rich regions of the membrane may facilitate and spread fatty acid peroxidation (see Chapter 20). That is, polyunsaturated phospholipids concentrated in fluid regions around lipid rafts may be especially sensitive to fatty acid peroxidation.
- Lipid whiskers not only act as signaling molecules activating phagocytic cells, but also likely significantly perturb other membrane functions, including proton and sodium permeability.

Data supporting these concepts can be found in papers reviewed by Catalá (2012).

13.2 LIPID WHISKERS SIGNAL PHAGOCYTIC CELLS TO CONVERGE ON A DAMAGED MEMBRANE SITE

Li and colleagues (2007) have proposed that receptors located on the surface of phagocytic cells seem to sense a lipid molecule on the membrane surface as a signal (Figure 13.1). This was the first clue that novel conformations of oxidized lipids might protrude into the aqueous phase. Thus, lipid whiskers are proposed to send signals to phagocytic cells to begin the degradation of any oxidatively damaged membrane surface deemed dysfunctional and a threat to neighboring cells— a toxic membrane state. The term *toxic membrane* requires clarification. Recall that cellular suicide (apoptosis) and phagocytosis are continuous and normal parts of the cell cycle. Oxidatively truncated, unsaturated phospholipids forming lipid whiskers that invite phagocytosis can be viewed as an important part of the overall defense network against oxidative damage. That is, cells that have sustained severe oxidative damage of their membranes are targeted for removal. We suggest that such an elaborate system of handling damaged membranes has evolved in part as a defense against oxidative stress in addition to energy stress. We also suggest that an excess of lipid whiskers is a sign of a cell or mitochondrion whose energy homeostasis is irreversibly dysfunctional, in essence activating the phagocytic process and destroying the unhealthy cell or organelle.

The authors of the lipid whisker model point out that the molecular architecture of membranes and the conformational dynamics of individual phospholipids are likely far more complex and important than previously anticipated. This led them to propose that the lipid whisker model might be a widespread mechanism encompassing critical events in aging, senescence, inflammation, and apoptosis.

A recent test of the lipid whisker model has been performed using receptor proteins on the membrane surface of macrophages as a probe to calibrate critical regions of putative lipid whiskers acting as signaling molecules (Gao et al., 2010). According to the 2008 rendition of the lipid whisker model (Greenberg et al., 2008; Hazen, 2008), receptors on phagocytes gain access to and bind specifically to the negatively charged head group of phospholipids, as well as recognizing the terminal carboxyl group of the truncated chain, the latter also negatively charged. Studies on the structural basis for the recognition of oxidized phospholipids by Gao and colleagues (2010) show that a high-affinity scavenger receptor CD36 recognizes and binds to three different chemical groups of truncated phospholipids—the long sn-1 hydrophobic (e.g., saturated) chain, the sn-3 hydrophilic phosphocholine or phosphatidic acid group, and the polar-truncated sn-2 tail. A negatively charged carboxylate at the terminus of the truncated chain satisfies binding at this location. All three sites are essential for high-affinity binding.

An interesting biochemical question concerns how receptor CD36 recognizes or binds to a long-chain, saturated fatty acid such as a stearic acid (18:0) chain embedded fairly deeply in the membrane. At least three mechanisms can be envisioned. A hydrophobic loop or region of the receptor protein might penetrate deeply enough into the interior of the outer membrane leaflet to recognize the C-18 chain. In another scenario, once the receptor makes initial contact with the lipid whisker, then the long acyl chain acting as anchor might be dragged out of the membrane into the aqueous phase by the much larger phagocytic cell. A third alternative is that the rate of flip-flop of the truncated phospholipid might be accelerated once oxidative cleavage occurs, exposing binding sites normally buried in the membrane. Note that lipid whiskers formed from DHA phospholipids usually occur on the cytosolic leaflet of the membrane. In senescing cells, scramblase enzymes might flip the defective phospholipid to the outer leaflet, where it is recognized by its receptor.

13.3   PHAGOCYTIC CELLS HYPERACTIVATED BY LIPID WHISKERS CAN INCREASE OXIDATIVE DAMAGE FOSTERING INFLAMMATION

A case history is chosen to illustrate how signaling by lipid whiskers can generate hyperactive phagocytes. Microglia are an amazing class of phagocytic cells that roam throughout the brain in search of pathogens, changes of homeostasis, or structural defects in neurons. Since the brain largely lacks white blood cells, microglia play this role, but are more generalized in responding to a plethora of signals besides invading pathogens. Microglia have numerous receptors on their ever-changing processes that allow them to constantly monitor chemical changes in the environment and probe the surfaces of neurons for defects. Microglia not only sense their environment, but also can also adapt quickly to counteract imbalances or structural defects with the goal of restoring homeostasis. For a vivid description of the versatile powers of microglia, see the article by Hanisch and Kettenmann (2007).

Truncated oxidation products of DHA chains in phospholipids are believed to flip to the membrane surface, where this DHA derivative acts as a signaling molecule for triggering phagocytosis by microglia. Microglia target the oxidized membrane surface of neurons via the scavenger receptor CD36, which locks onto the lipid whisker and triggers phagocytosis. In this manner, senescing axons, membrane fragments, and spent synaptosomes are removed. Microglia also engulf amyloid-beta plaque, the signature peptide of Alzheimer's disease, and have receptors for this tangle of toxic peptides. As signal strength through the CD36 scavenger receptor intensifies and is perhaps reinforced by parallel signaling pathways, microglia adapt by becoming increasingly active, entering a gray zone between being beneficial and harmful. Harmful in this case is defined as being too aggressive or hyperactive, to the point of initiating a state of sterile inflammation. Sterile inflammation contrasts with pathogen-induced inflammation, which is induced to kill the harmful invader. In the unfortunate case of sterile inflammation, hyperactive microglia create an inflammatory environment around neurons and destroy healthy synapses and axons, and eventually trigger neuron apoptosis.

Recently, Stewart et al. (2010) defined in molecular terms how the CD36 receptor promotes sterile inflammation. It is established that in Alzheimer's disease, deposition of amyloid-beta and oxidized lipids triggers a protracted sterile inflammation response. Although chronic stimulation of microglia was believed to underlie the pathology of Alzheimer's disease, the molecular mechanisms of activation were unclear. In this seminal study, Stewart and colleagues show that the scavenger receptor CD36, which receives signals from amyloid-beta and oxidized DHA phospholipids, stimulates sterile inflammation. This and other data from this laboratory define a molecular cascade of events, beginning with polyunsaturated membranes and leading to sterile inflammation. Thus, oxidation of polyunsaturated phospholipids has the potential to inflict damage on healthy neurons by hyperactivation of phagocytic cells.

Oxidatively damaged, senescing, or unhealthy nerve cells secrete a chemical messenger that induces the microglia to concentrate around focal points of neurodegeneration. The chemokine fractalkine, which docks onto a receptor protein on the surface of the microglial cells, is the likely signaling molecule. When this receptor was genetically eliminated in mice, nerve cell loss was prevented (Fuhrmann et al., 2010). These data show that stressed nerve cells secrete a chemical messenger that attracts microglia. Unfortunately, when large numbers of microglia gather around neurons, the ensuing inflammatory reaction can become too strong, resulting in the elimination of healthy axons and neurons. This implies that chemical signaling between nerve cells and microglia plays an important role in mediating neuron loss during the course of neurodegeneration. The current picture is that a battery of chemical signals are exchanged between neurons and microglia. DHA-mediated lipid whiskers fall into the category of signaling molecules between senescing neuron membranes and receptors on microglia (Silverstein, 2009). Pathogen-destroying white blood cells circulating in the rest of the body play roles similar to microglia, again contributing essential protection against pathogens balanced against oxidative risks caused by hyperactive white cells.

13.4 OXIDATIVELY TRUNCATED PHOSPHOLIPIDS AS TRIGGERS OF APOPTOSIS MEDIATED BY MITOCHONDRIA

T. M. McIntyre is a pioneer in the field of bioactivities of oxidatively truncated phospholipids, and he recently published a unified model of molecular biology of this class of lipids (McIntyre, 2012). We suggest his model linking oxidative stress, membrane unsaturation, oxidatively truncated phospholipids, energy uncoupling of mitochondria, and apoptosis goes a long way toward answering the famous chicken–egg scenario of which comes first—oxidative stress or energy stress? According to his model, oxidative stress is amplified at the level of dysfunctional mitochondrial membranes as energy stress triggering apoptosis (McIntyre, 2012). In his model, conventional polyunsaturated phospholipids, in contrast to cardiolipin, are the major sources of oxidatively truncated phospholipids. These molecules can leave their normal position in the membrane and move by diffusion through the aqueous phase, be transported across membranes, or even be concentrated in mitochondria, where they trigger energy uncoupling that can lead to apoptosis. The individual steps of this cascade triggering apoptosis are discussed in detail by McIntyre (2012).

The molecular mechanism of how truncated phospholipids damage mitochondrial structure and function is not fully elucidated but can be defined as follows:

- Hydrophilic truncated phospholipids interact directly with mitochondrial membranes.
- Uncoupling of the mitochondrial electrochemical proton gradient occurs.
- Energy uncoupling is reversible since sequestering of truncated phospholipids by albumin restores mitochondrial function.
- Intercalation of truncated phospholipids into mitochondria causes swelling mediated through the mitochondrial permeability pore (MPP).
- Evidence for interaction of truncated phospholipids at the level of MPP is based on the finding that cyclosporin A interferes with phospholipid-induced loss of the mitochondrial proton gradient.

The main take-home lesson from these data is that phospholipid oxidation products physically interact with mitochondria, causing some continuous energy wastage. McIntyre (2012) reviews data with model animals that suggest that the pathological effects of energy uncoupling caused by truncated phospholipids might become more severe with aging.

13.5   SUMMARY

Clearance of truncated phospholipids from the blood is extremely rapid (less than a thirty-second half-life) and involves uptake by the liver and kidney. The evolution and biochemistry of phospholipases that handle truncated phospholipids have been reviewed by McIntyre (2012). He concludes that oxidized phospholipids are rapidly and specifically hydrolyzed by phospholipases A_2, and that such hydrolases originally evolved to protect cells from oxidative cell death. We suggest that there is another perspective to these data that help illuminate its importance in understanding aging and age-dependent diseases. The main point is that enzymatic detoxification of oxidatively truncated phospholipids protects cells from death by both oxidative stress and energy stress. That is, rapid enzymatic clearance of truncated phospholipids from the bloodstream saves energy and alleviates energy stress because proton electrochemical gradients of mitochondria are conserved. In a long-lived mammal such as a human, energy gain by this mechanism might be considerable. Recall that energy conserved equates to more energy produced. Thus, if McIntyre's model of apoptosis is correct, then oxidative stress → energy stress → apoptosis, supporting the view that energy stress might be the "mother of all stresses" causing aging and age-dependent diseases.

REFERENCES

Catalá, A. 2012. Lipid peroxidation modifies the picture of membranes from the "fluid mosaic model" to the "lipid whisker model." *Biochimie* 94:101–9.

Farooqui, A. A., H.-S. Yang, T. A. Rosenberger, and L. A. Horrocks. 1997. Phospholipase A2 and its role in brain tissue. *J. Neurochem.* 69:889–901.

Fuhrmann, M., T. Bittner, C. K. Jung, et al. 2010. Microglial Cx3cr1 knockout prevents neuron loss in a mouse model of Alzheimer's disease. *Nat. Neurosci.* 13:411–13.

Gao, D., M. Z. Ashraf, N. S. Kar, et al. 2010. Structural basis for the recognition of oxidized phospholipids in oxidized low density lipoproteins by class B scavenger receptors CD36 and SR-BI. *J. Biol. Chem.* 285:4447–54.

Greenberg, M. E., X. M. Li, B. G. Gugiu, et al. 2008. The lipid whisker model of the structure of oxidized cell membranes. *J. Biol. Chem.* 283:2385–96.

Gugiu, B. G., C. A. Mesaros, M. Sun, et al. 2006. Identification of oxidatively truncated ethanolamine phospholipids in retina and their generation from polyunsaturated phosphatidylethanolamines. *Chem. Res. Toxicol.* 19:262–71.

Hanisch, U.-K., and H. Kettenmann. 2007. Microglia: active sensor and versatile effector cells in the normal and pathologic brain. *Nature Neurosci.* 10:1387–94.

Hazen, S. L. 2008. Oxidized phospholipids as endogenous pattern recognition ligands in innate immunity. *J. Biol. Chem.* 283:15527–31.

Leslie, C. C. 1997. Properties and regulation of cytosolic phospholipase A2. *J. Biol. Chem.* 272:16709–12.

Li, X. M., R. G. Salomon, J. Qin, et al. 2007. Conformation of an endogenous ligand in a membrane bilayer for the macrophage scavenger receptor CD36. *Biochemistry* 46:5009–17.

McIntyre, T. M. 2012. Bioactive oxidatively truncated phospholipids in inflammation and apoptosis: formation, targets, and inactivation. *Biochim. Biophys. Acta* 1818:2456–64.

Reis, A., and C. M. Spickett. 2012. Chemistry of phospholipid oxidation. *Biochim. Biophys. Acta* 1818:2374–87.

Silverstein, R. L. 2009. Type 2 scavenger receptor CD36 in platelet activation: the role of hyperlipemia and oxidative stress. *Clin. Lipidol.* 4:767.

Stewart, C. R., L. M. Stuart, K. Wilkinson, et al. 2010. CD36 ligands promote sterile inflammation through assembly of a Toll-like receptor 4 and 6 heterodimer. *Nat. Immunol.* 11:155–61.

14 Selective Targeting of HUFAs Away from Cardiolipin and Beta-Oxidation Combine to Protect Mitochondrial Membranes against Oxidative Damage

Selective fatty acid-targeting mechanisms allow human cells to incorporate specific fatty acids into specific membranes, thus maximizing benefits and avoiding risks. The biochemical basis of fatty acid targeting is poorly understood, but it is clear from compositional analysis that powerful mechanisms are at work to shuttle fatty acids to specific membranes or membrane domains. One of the best known examples is the targeting of docosahexaenoic acid (DHA) to the tail membrane domain of sperm (Chapter 12), which not only is enriched with DHA, but also houses a set of proteins specific for the tail domain. We suggest that DHA targeting has several advantages for sperm, including enhancing biochemical efficiency of motility and avoiding oxidative damage to germ-line DNA packaged in the head (Valentine and Valentine, 2009). In addition to sperm, humans target DHA to neurons of the brain and nervous system, with the highest levels found in the retina. In contrast, membranes of most human cells and mitochondria contain only traces of DHA and relatively low levels of other highly unsaturated fatty acids (HUFAs).

The human mitochondrial membrane is another example of a specialized bilayer. A sophisticated fatty acid-targeting system enriches cardiolipin in these membranes with polyunsaturated fatty acid tails, primarily 18:2. See Chapter 17 for a description of fatty acid uptake and biosynthesis in mitochondria, including the production of $(18:2)_4$-cardiolipin, a molecular species shown to be essential in humans (Chapter 8). The absence of DHA-cardiolipin in human mitochondrial membranes can be explained by a selective targeting system with the specificity to direct DHA away from mitochondrial membranes.

Recent data from tracer studies show that most of the DHA consumed in the human diet is degraded by beta-oxidation. We suggest that beta-oxidation not only is a major source of energy, but also acts as a regulatory system that maintains a low pool level of DHA. We further suggest that maintaining a low DHA pool size limits availability of DHA for incorporation into membranes. Note that beta-oxidation is shut down in the adult brain, where the blood-brain barrier strictly limits influx and efflux of DHA.

14.1 SELECTIVE FATTY ACID TARGETING HAS BEEN DEMONSTRATED FOR DHA IN SPERM AND OTHER CELLS

This section focuses on DHA-enriched lipid domains in sperm cell membranes. A membrane surrounds the entire sperm cell, and over 90 percent of DHA in monkey sperm is localized in the tail (Connor et al., 1998). The membrane of a sperm cell is divided into three distinct domains—anterior head, posterior head, and tail (Figure 14.1). This membrane domain structure was determined by immunofluorescence microscopy using a variety of monoclonal antibodies targeting specific proteins (antigens) on the sperm membrane surface. In some cases sperm membrane proteins are able to diffuse within the confines of their own domains, and it is believed that DHA follows a similar pattern. The barrier preventing DHA from leaving one domain for another is not fully understood. In addition to sperm, epithelial cells, which are among the most numerous cells in the body, are able to confine

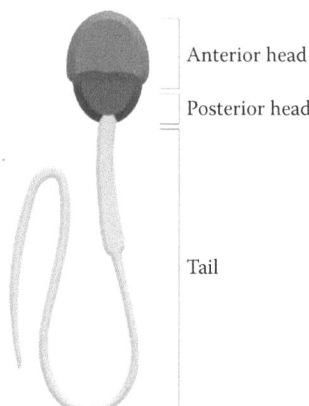

Anterior head

Posterior head

Tail

FIGURE 14.1 Membrane lipid domains allow cells to maximize benefits and reduce risks of unsaturated fatty acids. As shown in this figure, DHA is targeted largely to the tail lipid domain versus the head lipid domain of monkey sperm. Sperm have a third lipid domain surrounding the midpiece. Other cells and organelles, including neurons, epithelial cells, and mitochondria, have evolved selective fatty acid-targeting mechanisms resulting in generation of specialized membrane domains enriched with specific unsaturated fatty acids. Note that DHA is usually absent in human mitochondrial membranes, which are populated with linoleic acid (18:2). (From Alberts et al., *Molecular Biology of the Cell*, 5th ed., 2007. Garland Science-Books, NY. Copyright © 2007. Republished with permission of Garland Science-Books; permission conveyed through Copyright Clearance Center, Inc.)

proteins and lipids to specific membrane domains. Epithelial cells, such as those forming the surface of the colon, have evolved membrane domains called the apical plasma membrane facing the colon environment and the basal plasma membrane facing interior layers of cells. This membrane domain structure may allow epithelial cells to regulate membrane peroxidation levels through the localization of unsaturated fatty acids being incorporated and maintained in specific domains. For example, membrane fatty acids localized in the apical plasma membrane might be more or less prone to higher rates of peroxidation than interior phospholipids because of the exposure of these membranes to the colon environment. Senescing colonic cells, which are in direct contact with colon contents, appear to target DHA to cardiolipin as a mechanism to help assassinate these potentially precancerous cells (see Valentine and Valentine, 2009, for review and list of references).

Like epithelial cells, neurons have two distinct membrane domains, with the plasma membrane of the nerve cell body and dendrites resembling the basolateral membrane of an epithelial cell, while the plasma membrane of the axon and its nerve terminals correspond to the apical membrane. It is clear that specific proteins are localized in the membrane domains of neurons, and as in the case of sperm DHA, other specific fatty acids might also be localized into specific lipid domains.

Synaptic vesicles are a specialized class of tiny (~50 nm diameter) secretory vesicles that are unique not only in their small size and vast numbers, but also in their mechanism of origin and their high DHA and cholesterol content. Synaptic vesicles deliver neurotransmitters at synapses, and the importance of DHA in the biochemistry of the synaptosome cycle is reviewed in an earlier book (Valentine and Valentine, 2013). Synaptic vesicles are initially generated from the trans Golgi network and delivered to the plasma membrane of a synapse. During synaptic activity, the pool of synaptic vesicles, which may reach numbers as high as several thousand per synapse, can be directly retrieved from the plasma membrane of a synapse by endocytosis. The lipid content of synaptic vesicles is unique in at least two ways—extraordinarily high levels of both DHA and cholesterol. From a functional perspective this unusual lipid composition can be accounted for by (1) the cholesterol helping form a robust permeability barrier against proton leakage, and (2) DHA is proposed to contribute dual functions, including extraordinary membrane motion, to maximize the rate of synaptic cycling (see Chapter 4) while providing, in combination with cholesterol, a suitable permeability defense against proton leakage, presumably caused by extreme membrane curvature stress. The need for DHA for biochemistry in synaptic vesicles exposes DHA-enriched membranes of synapses to peroxidation. How DHA levels are maintained during cycling of synaptic vesicles is unknown. According to one scenario, the trans Golgi apparatus might initially target DHA to synaptic vesicles. During synaptosome cycling DHA might be selectively captured from the synaptic membrane surface by rapid retrieval by endocytosis. We propose that the net effect of localization of fatty acids into specific membrane domains maximizes benefit and reduces oxidative risk. According to this view, targeting DHA away from mitochondrial membranes might decrease both energy stress and oxidative stress.

14.2   SELECTIVE TARGETING AND AVOIDANCE OF POLYUNSATURATED CHAINS IN CARDIOLIPIN AS A PRO-LONGEVITY MECHANISM

In developing the central theme for this book (i.e., unsaturated membranes' contributions to aging), we sought some clues explaining why DHA is missing from mitochondrial membranes. The DHA principle (Valentine and Valentine, 2009) seems to explain the absence of DHA, leading us to question if the DHA principle could help explain in biochemical terms the extreme (thirty-fold) difference in longevity between a mouse and a human. One needs only to assume that the long life span of humans has some significant selective advantage to our species (see Chapter 7) to arrive at the starting point of a compelling concept. Perhaps humans, in contrast to mice, have evolved unique biochemical mechanisms aimed at increasing life span by somehow maximizing the benefits versus risks of DHA in mitochondria. We have already discussed the concept of selective DHA targeting, in which specific membranes such as synaptic vesicles or axons are enriched with DHA, while 18:2 is the predominant unsaturated fatty acid in mitochondrial membranes in the same cell. In a previous book (Valentine and Valentine, 2013) we suggested that human neurons might target DHA away from their mitochondria as a mechanism to avoid oxidative damage and enhance longevity of neurons, ultimately increasing brain span by decreasing the probability of neurodegeneration. Due to their evolutionarily honed short life span, mice would gain little from selective DHA targeting and might even be harmed. Indeed, DHA enrichment of mitochondrial membranes of mice seems to be necessary to turbocharge energy production to keep up with the hyperactive metabolism of this tiny animal (Hagopian et al., 2010; Chapter 11). The DHA content of membranes of heart cells in mice is about 10 to 15 percent of total fatty acids compared to about one-tenth this level of DHA in human cardiac cells. Thus, the DHA level in the heart of a mouse is about an order of magnitude greater than that of a human, in contrast to neurons of mouse versus human, in which DHA levels are high in both.

Data on the comparative DHA composition of membranes of mouse versus man support a selective DHA-targeting model. For example, in contrast to humans, mice target DHA and other HUFAs to their mitochondria. We interpret high DHA levels in mitochondrial cardiolipin as an indication of energy stress in a cell, and vice versa. This idea is consistent with data on the evolution of extreme flight in hummingbirds, where DHA levels in phospholipids are high and life spans are low (Chapter 6). In humans with long life spans, DHA is rarely incorporated into cardiolipin, and when it is, DHA-cardiolipin is generally considered to be a pathological or purposeful cellular suicidal molecule. Even in the absence of DHA highly polyunsaturated human mitochondria still retain a good bit of conformational flexibility of their inner membranes, a property essential for energy production (Chapter 10). The most common unsaturated molecular species in humans is $(18:2)_4$-CL. For example, $(18:2)_4$-CL is the major molecular species in human heart muscle, whose mitochondria are considered to be relatively hard working. Heart rates increase dramatically in tiny mammals, reaching greater than 1000 beats per minute in miniature voles. DHA enrichment of mitochondrial membranes of cardiac cells is known to increase proportionally with heartbeat (Chapter 6). Historically, comparative data on the direct relationship between DHA levels and heart rates first alerted scientists of a possible inverse relationship between DHA content of membranes and life span and led to the membrane pacemaker theory of aging (Chapter 6).

According to the membrane unsaturation hypothesis of aging, targeting DHA randomly into all membranes of human cells as occurs in mice would be predicted to be pro-aging, exerting dramatic downward pressure on both life span and brain span. Instead, human neurons appear to target DHA away from cardiolipin (Yabuuchi and O'Brien, 1968; Söderberg et al., 1992; Kirkland et al., 2002; Kiebish et al., 2008, 2009a, 2009b), a finding that is consistent with the absence of DHA-cardiolipin in most human mitochondria. This distribution pattern can be explained biochemically by the two-stage process mitochondria use to synthesize cardiolipin (Chapters 5 and 17). In the first stage, cardiolipin synthetase located specifically in mitochondria joins two phospholipids together through their head groups, generating a relatively saturated species of cardiolipin. The second stage involves a remodeling process that adds increasing levels of unsaturated chains to cardiolipin. Specificity mechanisms ensure that the molecular species $(18:2)_4$-CL is the most abundant of the highly polyunsaturated molecular species of CL in human mitochondria. The presence of DHA and other HUFAs in CL suggests that the second stage of CL biosynthesis is far more active in mice than in humans.

We suggest that mouse mitochondrial bioenergetics has evolved along a different track from that of humans. One idea is that bioenergetic gain is more important in mice than longevity. Mitochondria in mice are expected to undergo more cycles of replication and acquire mtDNA mutations at a faster rate than human mitochondria. In humans it is proposed that targeting DHA away from mitochondrial membranes goes a long way toward increasing life span. However, specialized cells, including neurons, still depend on DHA as an essential membrane building block for axons and synaptic vesicles, but seem to maximize its benefits by avoiding incorporation into mitochondria. Selective targeting of DHA away from mitochondria is regarded as a mechanism to ensure longevity, especially of the human brain, but requires that such targeting mechanisms remain functional for a lifetime. For a recent review of the literature linking cardiolipin and neurodegeneration, see Paradies et al. (2011). Also see the article by Gruber and colleagues (2011) for data described in Chapter 4 on the role of CL peroxidation as a trigger and reporter for aging in *Caenorhabditis elegans*. The role of cardiolipin as a trigger of programmed cellular death and mechanisms to protect CL are described in Chapter 20.

14.3   BETA-OXIDATION IS RESPONSIBLE FOR DEGRADING A VAST MAJORITY OF DHA IN THE BODY, THUS MINIMIZING DHA INCORPORATION INTO MOST CELLULAR MEMBRANES

Humans have evolved a powerful fatty acid oxidation (beta-oxidation) system operating in many organs for degrading energy-rich fatty acids, including DHA, which are catabolized as a major source of cellular energy (Figure 14.2). In contrast, the adult brain containing the largest amounts of DHA-enriched membranes in the body does

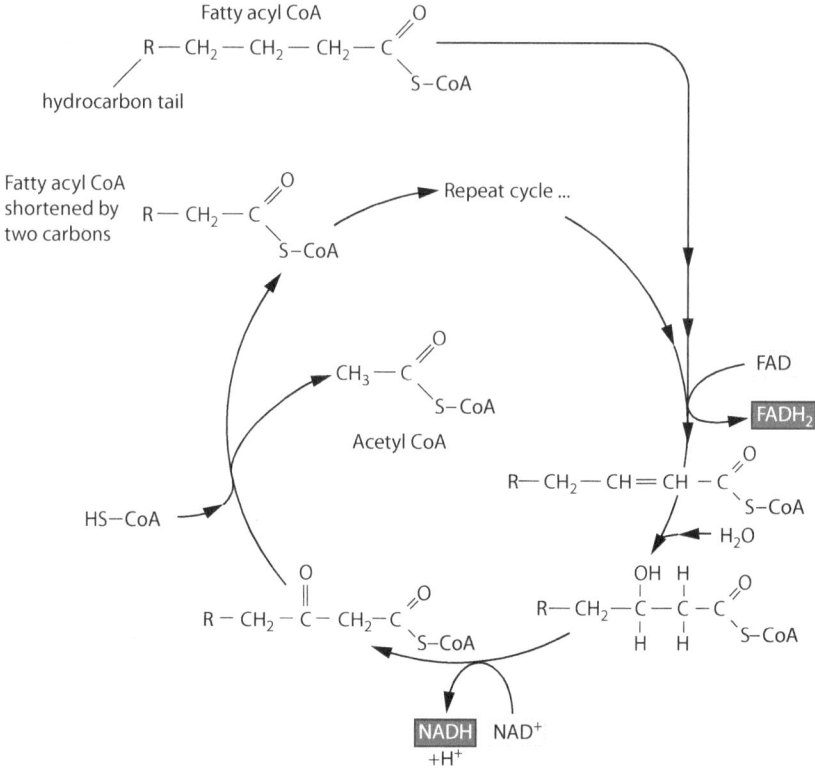

FIGURE 14.2 The fatty acid oxidation cycle. The cycle is catalyzed by a series of four enzymes in the mitochondrion. Each turn of the cycle shortens the fatty acid chain by two carbons and generates one molecule of acetyl CoA and one molecule each of NADH and FADH$_2$. (From Alberts et al., *Molecular Biology of the Cell*, 5th ed., 2007. Garland Science-Books, NY. Copyright © 2007. Republished with permission of Garland Science-Books; permission conveyed through Copyright Clearance Center, Inc.)

not carry out beta-oxidation of DHA or other fatty acids. Thus, once DHA passes across the blood-brain barrier, which severely limits trafficking of DHA in and out of the brain, DHA appears to be surprisingly stable (see Chapter 15). However, many other organs, especially the liver, readily degrade fatty acids not only as sources of energy, but also as a source of carbon skeletons for biosynthesis. The net effect is that individuals who consume significant amounts of seafood end up degrading the vast majority of dietary DHA as energy food. Thus, beta-oxidation plays a secondary but key role in regulating body-wide levels and availability of DHA for membrane biosynthesis.

A detailed description of the biochemistry or enzymology of beta-oxidation is beyond the scope of this book, and instead we focus on some specific aspects of DHA breakdown by humans. A recent tracer study using ^{13}C-DHA fed to human volunteers led to the following conclusions regarding the time course and importance of beta-oxidation of DHA (Plourde et al., 2011):

- Beta-oxidation of ^{13}C-DHA peaks at four hours postintake.
- About 35 percent of total DHA is oxidized after seven days.
- In about three weeks most of the labeled DHA is degraded and respired as CO_2.
- For comparison, rates of beta-oxidation of ^{13}C-alpha-linolenic acid, the major precursor for biosynthesis of DHA, are about twice as fast as those of DHA.

The data above reflect activity of beta-oxidation of DHA in all parts of the body except the brain, where DHA homeostasis and rates of turnover are dramatically different. As already mentioned, beta-oxidation of DHA and other fatty acids in brain tissue is shut down. Recent data, discussed in Chapter 15, show a surprisingly different picture of DHA turnover in the human brain compared to the rest of the body (Umhau et al., 2009).

Data from the tracer experiment discussed above show that beta-oxidation plays a central role in degrading much of the combined dietary and biosynthesized DHA in the body. The brain needs only a tiny fraction of the total DHA circulating in the bloodstream, and after passage across the blood-brain barrier, most of the DHA somehow becomes highly protected. However, DHA molecules in the rest of the body are subject to beta-oxidation, which has the effect of decreasing DHA levels available for membrane synthesis in most human cells. In the absence of dietary DHA, biosynthesis in the liver and other organs is able to meet much of the demand for DHA needed for synthesis of specialized membranes in adults. Alpha-linoleic acid (18:3 n-3), the major precursor of DHA biosynthesis, is often available in oils derived from plants. Thus, humans synthesize DHA starting with building blocks normally found in excess in our diet. Beta-oxidation also strongly modulates 18:3 n-3 levels in the body, helping maintain this key precursor to DHA at relatively low levels, in tune with the relatively small amount of DHA needed by the adult brain.

In summarizing this section, beta-oxidation is often overlooked as a major mechanism for maintaining a low pool size of DHA in the body. Recent tracer studies of DHA in humans are consistent with beta-oxidation being one of the most important mechanisms for maintaining low levels of DHA in the body.

14.4 COMPARATIVE BIOCHEMISTRY OF DHA DETOXIFICATION

The phrase DHA detoxification is a new concept that requires further definition. The first point is that DHA (or eicosapentaenoic acid (EPA)) detoxification mechanisms have evolved across a wide range of organisms and beta-oxidation in only one of them, as follows:

- *Escherichia coli*: DHA appears to be toxic, and a biochemical block at the level of DHA uptake seems to prevent incorporation in membranes. Note that a hierarchy of unsaturated fatty acids with one to five double bonds is taken up and incorporated into membrane phospholipids by *E. coli*.
- *Shewanella*: DHA → EPA retroconversion.
- *C. elegans*: DHA → EPA retroconversion.

- Krill: Enzymatic detoxification of excessive dietary EPA in the mouth and gizzard by activation of a powerful lipoxygenase system.
- Baleen whales: Detoxification via hydrogenation by symbionts in the rumen-like forestomach.
- Humans: Beta-oxidation.

It is interesting to compare the mechanisms of DHA/EPA detoxification between humans and whales as examples of mammals with maximum life spans greater than a century. It is clear that whales live as long or longer than humans, but the intriguing claim that a bowhead whale reached an age of 211 years requires further verification (Chapter 6). Note that blue whales known to live more than a century can consume an estimated 70 pounds of DHA and other HUFAs per day in their diet of krill. The brain size of whales is around 17 to 20 pounds—about six times that of humans. It can be estimated based on the levels of DHA needed by the human brain (Chapter 15) that a whale requires about 25 mg per day of DHA to maintain DHA homeostasis in its brain. Thus, a blue whale is estimated to consume around 1500 times more DHA than needed to maintain homeostasis of DHA in the brain. Whale meat is lean and contains only a fraction of 1 percent of DHA, a proportion remarkably different than that of krill. Blubber contains DHA, but again, levels are far below those of krill. Levels of EPA and other HUFAs follow a similar distribution pattern in the whale body, with the exception of the brain, in which EPA levels are at least ten-fold lower than those of DHA. Whereas it is not possible to derive a detailed balance sheet for DHA in whales, we suggest that whales, like other ruminant animals, detoxify an excess of dietary DHA, EPA, and other HUFAs by hydrogenation in the forestomach (Chapter 6). When whales sleep or carry out deep dives, unlike humans, they are able to shut off one side of the brain at a time. It is interesting to speculate that during sleep, the brain respiration rates and membrane peroxidation rates plummet—yet another mechanism to protect mitochondrial membranes. Thus, hummingbirds, humans (Chapter 15), and whales might rely on O_2 avoidance to protect their mitochondria.

Obviously, the average human diet contains a miniscule amount of DHA compared to that of a whale, but it is estimated that individuals who routinely eat fish consume far more DHA than the few milligrams per day needed to maintain DHA homeostasis in the adult brain. As described above, beta-oxidation of DHA not only decreases overall DHA levels, but also releases a considerable amount of energy to the cell. Thus, beta-oxidation in humans appears to be a major mechanism to detoxify an excess of DHA.

14.5 SUMMARY

Selective targeting of fatty acids and beta-oxidation work together to protect mitochondrial membrane lipids, especially cardiolipin, against peroxidation. Selective fatty acid targeting allows cells to place specific fatty acids in membranes or membrane domains where oxidative risks to the whole organism are minimized and benefits are maximized. The evolution of specialized membranes, including rhodopsin disks that are exquisitely sensitive to oxidative damage, seems to defy this rule. However, vision is the paramount sensory system in humans, and benefits clearly outweigh oxidative risks, which are handled by multiple defenses, including phagocytic recycling of spent disks and minting of new disks (Chapter 12). Selective targeting of DHA away from most human cells protects a majority of cells from oxidative damage caused by DHA. This same concept applies to mitochondrial membranes. Selective fatty acid targeting is understudied and deserves more attention (Pfenniger, 2009).

Beta-oxidation is well known as a powerful biochemical pathway to extract energy from fats and supply the cell with carbon skeletons needed for growth. We suggest that beta-oxidation also helps govern the pool sizes of unsaturated fatty acids essential as membrane building blocks. For example, levels of DHA circulating in the bloodstream are governed by beta-oxidation, in essence limiting the amount of DHA available for incorporation into membrane phospholipids.

REFERENCES

Connor, W. E., D. S. Lin, D. P. Wolfe, and M. Alexander. 1998. Uneven distribution of desmosterol and docosahexaenoic acid in the heads and tails of monkey sperm. *J. Lipid Res.* 39:1404–11.

Gruber, J., L. F. Ng, S. Fong, et al. 2011. Mitochondrial changes in ageing *Caenorhabditis elegans*—what do we learn from superoxide dismutase knockouts? *PLoS One* 6:e19444.

Hagopian, K., K. L. Weber, D. T. Hwee, et al. 2010. Complex I-associated hydrogen peroxide production is decreased and electron transport chain enzyme activities are altered in n-3 enriched fat-1 mice. *PLoS One* 5(9):e12696.

Hulbert, A. J., R. Pamplona, R. Buffenstein, and W. A. Buttemer. 2007. Life and death: metabolic rate, membrane composition, and life span of animals. *Physiol. Rev.* 87:1175–213.

Kiebish, M. A., X. Han, H. Cheng, et al. 2008. Cardiolipin and electron transport chain abnormalities in mouse brain tumor mitochondria: lipidomic evidence supporting the Warburg theory of cancer. *J. Lipid Res.* 49:2545–56.

Kiebish, M. A., X. Han, H. Cheng, et al. 2009a. In vitro growth environment produces lipidomic and electron transport chain abnormalities in mitochondria from non-tumorigenic astrocytes and brain tumors. *ASN Neuro* 1(3):pii:e00011. doi: 10.1042/AN20090011.

Kiebish, M. A., X. Han, and T. N. Seyfried. 2009b. Examination of the brain mitochondrial lipidome using shotgun lipidomics. *Methods Mol. Biol.* 579:3–18.

Kirkland, R. A., R. M. Adibhatla, J. F. Hatcher, et al. 2002. Loss of cardiolipin and mitochondria during programmed neuronal death: evidence of a role for lipid peroxidation and autophagy. *Neuroscience* 115:587–602.

Paradies, G., G. Petrosillo, V. Paradies, et al. 2011. Mitochondrial dysfunction in brain aging: role of oxidative stress and cardiolipin. *Neurochem. Int.* 58:447–57.

Pfenniger, K. H. 2009. Plasma membrane expansion: a neuron's Herculean task. *Nat. Rev. Neurosci.* 10:251–61.

Plourde, M., R. Chouinard-Watkins, M. Vandal, et al. 2011. Plasma incorporation, apparent retroconversion and β-oxidation of 13C-docosahexaenoic acid in the elderly. *Nutr. Metab. (Lond.)* 8:5.

Söderberg, M., C. Edlund, I. Alafuzoff, K. Kristensson, and G. Dallner. 1992. Lipid composition in different regions of the brain in Alzheimer's disease/senile dementia of Alzheimer's type. *J. Neurochem.* 59:1646–53.

Umhau, J. C., W. Zhou, and R. E. Carson. 2009. Imaging incorporation of circulating docosa-hexaenoic acid into the human brain using positron emission tomography. *J. Lipid Res.* 50:1259–68.

Valentine, R. C., and D. L. Valentine. 2009. *Omega-3 fatty acids and the DHA principle*. Boca Raton, FL: Taylor and Francis Group.

Valentine, R. C., and D. L. Valentine. 2013. *Neurons and the DHA principle*. Boca Raton, FL: Taylor and Francis Group.

Yabuuchi, H., and J. S. O'Brien. 1968. Brain cardiolipin: isolation and fatty acid positions. *J. Neurochem.* 15:1383–90.

15 Oxygen Limitation Protects Mitochondrial Phospholipids, Especially Cardiolipin

The highest circulating oxygen levels in the human body occur in freshly oxygenated blood leaving the heart, while at the other extreme, the level of oxygen drops to zero in the anaerobic interior of the large intestine. Oxygen levels also fluctuate among organs and in different locations within organs, such as the brain (Figure 15.1) (Phelps, 2000; Kubicki et al., 2007), and especially in mitochondria due to their oxygen-scavenging ability. There are numerous mechanisms to explain why oxygen levels are so variable in different regions of the body, and it is clear that in many cases O_2 concentration is carefully and purposely regulated. Recent data suggest oxygen avoidance is widely practiced by many organisms as a key mechanism to protect their highly unsaturated membranes against oxidative damage (see Section III). Oxygen avoidance or limitation is an effective mechanism because it chokes off the supply of O_2 needed as substrate for membrane peroxidation.

Darwinian selection in humans has honed the levels of unsaturation in mitochondrial membranes to a minimum, but polyunsaturate-containing bilayers of mitochondria are still highly sensitive to peroxidation. We suggest that humans have evolved O_2 limitation as yet another important membrane-protective mechanism, as introduced in Chapter 12. Sperm cells avoid oxidative damage to their docosahexaenoic acid (DHA)-enriched tail membranes while traveling in the low-O_2 niche found in the female reproductive tract. Is this an isolated case, or do humans harness O_2 limitation on a broader scale as a membrane-protective mechanism? Recent data by Leedo and colleagues (2013) are consistent with differential regulation of O_2 levels in different regions of the brain as well as in separate locations within neurons.

15.1 DHA TURNOVER IN THE BRAIN IS SURPRISINGLY SLOW, SUGGESTING THE OPERATION OF NOVEL PROTECTIVE MECHANISMS

The DHA principle predicts that the brains of humans should set the pace of aging, but this clearly is not always the case since the brains of many individuals remain healthy through old age. However, the fact that about 50 percent of individuals display symptoms of neurodegeneration by age eighty-five is consistent with the concept

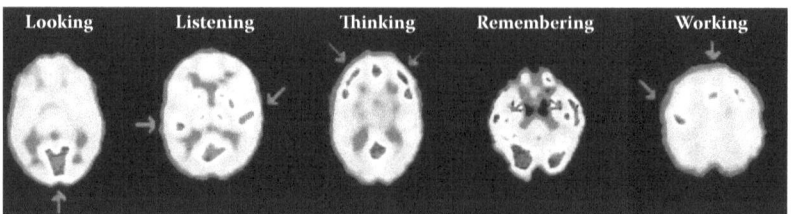

FIGURE 15.1   See color insert. Positron emission tomography (PET) shows that metabolic activity and respiration in different regions of the brain are activated by different mental tasks. Since glucose consumed as a major energy source must be tightly coupled to lactate utilization by mitochondria, these data suggest that oxygen consumption and levels are also variable. Although a robust circulatory system feeds large amounts of oxygen to the brain, vast numbers of neurons and astrocytes use much of the oxygen, with the overall effect being a differential lowering of oxygen levels throughout the brain. (From Phelps, *Proc. Natl. Acad. Sci. USA* 97:9226–33, 2000. Copyright © 2000, National Academy of Sciences, U.S.A.)

that the human brain often acts as a single organ, governing aging. We suggest that numerous specialized mechanisms, including oxygen avoidance, have evolved for long-term protection of the brain, enabling brain span with implications for aging human cells in general.

The following conditions in the human brain are predicted to favor fast rates of oxygen-mediated damage to membranes, especially mitochondria, axons, and synaptosomes:

- Massive blood supply for oxygenating the brain
- Warm temperature (i.e., body temperature of 37°C)
- Neural membranes enriched with DHA
- High levels of 18:2-CL in mitochondrial membranes

Sensitive chemical analytical procedures have been developed to detect the formation of oxidation products of DHA in the brain. F-4 neuroprostanes act as specific biomarkers for oxidative damage to DHA-enriched membranes of neurons (Dalle-Donne et al., 2006; Montine et al., 2002, 2005, 2007; Poon et al., 2004). The detection of F-4 neuroprostanes in cerebrospinal fluid validates the concept that at least some DHA chains in neuronal membranes are targets of oxidative damage. However, these data do not answer the important question regarding the rate of turnover of DHA in neurons.

Recently, another line of experiments has been developed with an eye toward calibrating the rates of turnover of DHA in the human brain. These experiments carried out at the National Institutes of Health involved injection of fourteen healthy volunteers with [1-^{11}C] DHA and following its fate and turnover in the brain using positron emission tomography (PET) (Umhau et al., 2009). The light isotope of carbon chosen for this experiment has been shown to be safe in human subjects. The data show that most of the DHA in neuron membranes is surprisingly stable, taking greater than five years for complete replacement of all DHA in the brain. However, after only forty-nine days, 5 percent of total DHA is estimated to be turned over, consistent with the idea that at least a certain fraction of DHA is rapidly degraded. For comparison with a short-lived mammal, DHA turnover in rat brain is roughly three times faster, with a 30 percent reduction in 105 days. Just one generation of omega-3 deprivation in rats led to behavioral changes (DeMar et al., 2006). It is of interest to compare the tracer method used by Umhau and colleagues (2009) to measure DHA turnover in the brain with the tracer method used by Young (1967) to calibrate what turned out to be a dramatically fast rate of turnover of rhodopsin membrane disks in rod cells (Chapter 12). In the latter case, radioactive amino acids were used to label rhodopsin protein embedded in newly minted rhodopsin membrane disks. In just ten days the complete stack of about 1000 rhodopsin membrane disks in rod cells was turned over. Roughly half of the DHA chains in senescing disks are damaged during a complete renewal cycle of ten days. The slow disappearance of total radioactive carbon from DHA molecules in the human brain is likely due to unexpectedly slow rates of chemical oxidative degradation of this fatty acid. Recall that enzymatic breakdown of DHA by beta-oxidation is absent in brain tissue. Thus, these data are consistent with membrane peroxidation as a major route of DHA destruction in the brain. The main point is that the slow turnover of DHA in brain membranes is a surprise, with complete turnover estimated to take five years, compared to ten days for loss of 50 percent of total DHA in rhodopsin disks.

There are a number of clues in the literature that suggest that the brain has evolved a regulated system of oxygen limitation, with major implications for understanding brain span. In a review of O_2 homeostasis in the brain, Ndubuizu and LaManna (2007) suggest that neurons of the brain cortex normally exist in a low-oxygen environment. They further suggest that when neurons become activated and require more oxygen, local capillary blood is increased in proportion to demand. This model is consistent with O_2 avoidance as a mechanism to protect neuronal membranes. We suggest that the apparently sophisticated regulation of O_2 levels in the brain evolved in part as a mechanism to protect not only neurons, but also mitochondrial membranes against oxidative damage, enabling brain span. How low do O_2 levels get in brain mitochondria or mitochondria of other tissues?

15.2 O_2 AVOIDANCE BY MITOCHONDRIA

The idea that mitochondria depend on oxygen avoidance as a defense against oxidative damage has been suggested by several different researchers during the past eighteen years (Skulachev, 1996; Erecińska and Silver, 2001; Turrens, 2003; Murphy, 2009). It is generally agreed by these authors that the intramitochondrial oxygen concentration is much lower than ambient (as a reference, 200 μM O_2 is present in air-saturated buffer solution at 37°C). The concentration of O_2 within mitochondria is a summation of many variables, with reported values ranging from 3 to 30 μM (Turrens, 2003).

Cristae membranes of mitochondria are specialized for the purpose of energy production consuming O_2 in the process. Mitochondrial membranes are also specialized with respect to their high enrichment with polyunsaturated fatty acids (PUFAs). We suggest that the inner membrane represents the predominant target of membrane peroxidation in human cells. This last statement brings us back to the central theme of this book—membranes' contribution to longevity. How do cristae membranes contribute to longevity? To gain perspective on this question, it is revealing to summarize the lipid dynamics of cristae membranes as follows (for a review, see Chazotte and Hackenbrock, 1988):

- Unsaturation levels of cristae (inner) membranes are much higher than those of the outer membrane.
- Electron transport complexes are localized in cristae membranes, with integral membrane proteins covering about 40 to 50 percent of the membrane surface.
- About 30 percent of membrane lipids are believed to be present in the boundary layer of membrane proteins, with the remainder present as naked membrane surface.
- Cristae membrane lipids are considered to be highly fluidizing, especially cardiolipin, in which 90 percent of fatty acid chains are unsaturated.
- Components of the respiratory chain have been calibrated to move laterally in cristae at relatively rapid rates.

These data led to the random collision model of mitochondrial electron transport (see Chazotte and Hackenbrock, 1988). In recent years collision theory has been overshadowed by data suggesting that cardiolipin acts as a molecular glue to generate supermolecular electron transport complexes for the purpose of increasing rates of respiration (Joshi et al., 2009). However, these data do not rule out the concept that collisions among electron transport components still play an important role in governing rates of respiration (see Chapter 10). The formation of supercomplexes of respiratory enzymes has also been suggested to be a mechanism against oxidative stress (Joshi et al., 2009). The localization and crowded conditions of electron transport components housed in cristae membranes raise another point related to oxidative stress. Recall that rates of peroxidation of membrane PUFAs increase linearly with increasing concentration of O_2, ranging from 0 to 95 percent O_2 (see Chapter 12). It is also clear that levels of O_2 modulate electron leakage from complex 1 (Chapter 16). O_2 sequestered by red cells carried in the bloodstream reaches cristae membranes after passing through, in sequence, the plasma membrane of the mother cell, cellular cytoplasm, outer mitochondrial membrane, intermembrane space, smooth portion of the inner membrane, and finally to cristae. Recall that the cristae junction pinches off the entrance into cristae, narrowing the opening to about 17 to 20 nm. Thus, the entry of O_2 directly from the intermembrane space might be hindered, further adding to the tortuous route of oxygen that is eventually scavenged by cytochrome oxidase (complex 4) of the electron transport chain.

Complex 4 of the respiratory chain acts as a terminal oxidase donating spent electrons from the electron transport chain to O_2 and forming water as a by-product of respiration. Complex 4 is a powerful scavenger of O_2 with strong binding affinity for O_2 as substrate. The overall effect is that complex 4 is able to carry out its essential role in supporting respiration at levels of O_2 much lower than is optimal for the chemical reactions involved in membrane peroxidation or ROS production by complex 1. Complex 4, as an untethered electron transport complex, moves laterally over the membrane surface, scavenging O_2 molecules as it goes. The random paths of all such mobile complex 4 molecules essentially blanket the membrane surface and often cross paths with complex 1. The O_2 scavenging power of complex 4 creates a microclimate of low O_2 around complex 1 and offers the benefits of decreasing ROS production, lowering rates of membrane peroxidation, and protecting membranes' proteins.

15.3 DURING O_2-LIMITED CONDITIONS CELLS HAVE THE OPTION OF REBALANCING THE RATIO OF RESPIRATION TO GLYCOLYSIS: CASE HISTORY OF SPERM

The case history of O_2 avoidance by sperm serves as an extreme example of how cells can modulate their ratios of respiration to glycolysis. As background, students of human fertility have been taught for generations that numerous mitochondria concentrated in a region just below the sperm head provide energy for flagellar whipping. During the past decade convincing data have led to a major revision of the conventional view of how sperm energize their flagella. The surprise is that mammalian sperm have evolved a novel bioenergetic adaptation to supply ATP for energizing their flagella while traveling in the low-O_2 environment of the female reproductive tract. Previously, mitochondria seemed to be the obvious choice for generating ATP for sperm motility. Instead, it has now been found that glycolytic enzymes spaced along the length of the tail cytoplasm, not mitochondria, are essential for fertility. That sperm cells produce lactic acid from glucose was established more than three decades ago, and appropriate biochemical activity for various glycolytic enzymes, including glucokinase, lactate dehydrogenase, and glyceraldehyde 3-phosphate dehydrogenase, has since been found in the tail region of a variety of mammalian sperm. It has also been established that hyperactivated motility is dependent on ATP from glycolysis and that inhibitors against mitochondrial ATP production do not block fertilization. These data overturn the conventional concept that mitochondrial bioenergetics provide the bulk of ATP for movement of sperm during the fertilization process. Formal proof as to the importance of glycolysis in energizing flagella came from data generated using knockout mutations of the essential glycolytic enzyme, glyceraldehyde dehydrogenase, in mice (Miki et al., 2004; Mukai and Okuno, 2004). Note that these data do not rule out an essential role of sperm mitochondria in providing ATP during sperm development in the testes or in other oxygenated environments.

The case of O_2 avoidance by sperm serves as a model for how localized O_2 avoidance might benefit neurons. As introduced above, the highly oxygenated environment characteristic of the brain would seem to be an unlikely place to look for O_2 avoidance mechanisms. However, the sperm model has led us to broaden our thinking to consider any mechanism that might lower O_2 levels in or around neurons or their parts, and protect membranes. We suggest that O_2 avoidance in axons of white matter might involve myelin as a physical barrier against O_2, with putative localized up-regulation of glycolysis in axons compensating for any deficiency in ATP due to low O_2 levels restricting respiration. According to this idea, the myelin sheath acts not only as insulation to increase efficiency and speed of neural impulses, but also as a physical barrier, allowing much of the outer surface area of myelinated axon membranes to limit their exposure to O_2. Obviously, such a barrier is not found at axonal internodal regions where O_2 is free to diffuse across the bare axon membrane. Membranes at synapses are not covered by myelin either, allowing this mitochondria-rich region to obtain a supply of O_2 from the environment. It is estimated that white matter is composed of roughly 100,000 miles of myelinated axons, and the point raised here is that the O_2 levels lying beneath the myelinated regions of axons might be subambient compared to the most oxygenated regions of the brain. This hypothesis can be generalized to include synapses and even axons of gray matter where active mitochondrial respiration in the head region, synaptic region, and dendrites might communally lower O_2 levels well below that of oxygenated blood or O_2 levels surrounding neurons. Also, the vast numbers of glial cells that support brain function communally carry out robust respiration and represent a major sink for O_2. Thus, glial cells have the respiratory power to lower O_2 levels of the brain as well as in their own cytoplasm and mitochondria.

Summarizing this section, the model of specialized O_2 avoidance in myelinated axons proposed here has two parts—myelin as an O_2 barrier combined with localized, up-regulated rates of glycolysis to compensate for possible lowered rates of energy production from respiration within the axon. In essence, what we are proposing is a mechanism similar to, but not as extreme as, that seen in the case of O_2 avoidance by sperm. Glucose is well known to act as the bulk energy source for the brain and is degraded by glycolytic enzymes yielding ATP and lactic acid. Obviously, a knockout of glycolysis in the brain would be lethal. Even though glycolysis yields only two ATP, along with two lactates per glucose molecule consumed, the rates of glycolysis can be extremely high. The point is that glycolysis should not be considered as second-rate or inferior to mitochondria for ATP production in axons, with the caveats that glucose is abundant and lactate homeostasis is strictly maintained by rapid mitochondrial respiration to avoid lactic acidosis. Mitochondria are responsible for complete oxidation of lactate, whose energy content supplies the bulk of the ATP needed by the brain. Numerous brain diseases in humans, including many mitochondria-mediated neurodegenerative disorders, are caused by dysfunction in lactic acid homeostasis in which lactic acid rises to toxic levels. A healthy brain maintains a tight linkage between glycolysis and respiration. We suggest that it might be possible in myelinated axons to strike a delicate balance in which glycolysis is moderately up-regulated to compensate for a decrease in energy from respiration caused by a deficiency of O_2 as substrate for mitochondria. This scheme satisfies the definition of specialized oxygen avoidance and is consistent with extraordinary stability of most of the DHA in the brain.

Several other possible mechanisms of O_2 avoidance for protecting membranes are worth considering for future research. For example, gray matter, which makes up less than half of the neuron mass of the brain compared to about 60 percent for white matter, consumes 94 percent of the oxygen. Does this mean that the robust rates of respiration by gray matter are harnessed to maintain low levels of O_2 in this critical brain region? The converse of this question applies to white matter. Is glycolysis more important for supplying energy to white matter than to gray matter? Is signal transmission (the central role carried out by white matter) less energy-intensive compared to mental processing carried out by gray matter, thus requiring less energy for functions of neurons in white matter? A point to consider is that different bioenergetic (Leedo et al., 2013) and O_2 avoidance mechanisms might have evolved to meet the needs of different key regions of the brain.

15.4   MODEL FOR O_2 PROTECTION BY MYELIN FOUND IN ROOT NODULE BACTERIA

A root nodule of a soybean plant contains about a billion or so bacteroids that convert nitrogen gas (N_2) into ammonium (NH_4^+) needed for plant growth. This essential process, called N_2 fixation, would not be possible without a sophisticated system of O_2 avoidance. The nitrogen fixation machinery, especially nitrogenase, is exquisitely sensitive to O_2. Oxygen not only denatures nitrogenase, but also pirates high-energy electrons needed to drive the reduction of N_2 (note that six high-energy electrons along with sixteen ATP are required as substrates for nitrogenase). Bacteroids face a dilemma because these cells energize symbiotic nitrogen fixation via oxygen respiration. This sets up a catch-22 in which O_2 is both essential and deadly with respect to symbiotic N_2 fixation. At least four major adaptations featuring O_2 avoidance are required to protect nitrogenase as follows:

- Leghemoglobin binds and delivers O_2 to bacteroids.
- Cytochrome oxidase of bacteroids operates at extremely low O_2 levels.
- Bacteroids enrich their membranes with a cholesterol-like molecule, which may hinder O_2 diffusion into the cytoplasm.
- Collective respiration of tightly packed bacteroids is expected to lower O_2 levels.

In essence, bacteroids, many of which are genetically dead, have earmarks of specialized mitochondria residing inside plant cells for the purpose of delivering not ATP, but rather ammonium to their host. Unlike mitochondria, which must protect their highly unsaturated membranes against oxidative damage, bacteroids are faced with protecting a highly O_2-sensitive enzyme system.

Perhaps the most remarkable mechanism to protect nitrogenase occurs in the symbiotic, nitrogen-fixing organism *Frankia* sp. (Berry et al., 1993). This actinomycete-like bacterium is able to induce nitrogenase when free-living cells are cultured with low oxygen. Under these conditions, normally elongated cells begin to differentiate into multicellular ovoid cells called vesicles, where N_2 fixation takes place. *Frankia* root nodules contain millions of vesicles, each of which is wrapped with up to fifty layers of a membrane composed of up to 87 percent terpenoids, a cholesterol-like molecule derived from squalene. The many membrane layers are proposed to block permeability of O_2, helping create an anaerobic environment that protects nitroge- nase. Another important nitrogen fixer, cyanobacteria, have evolved specialized, non-N_2-fixing cells called akinete, which seem to be protected from O_2 toxicity by a membrane insulation mechanism similar to *Frankia*. The main point is that these data validate the concept that wrapping cells with many layers of membranes enriched with cholesterol-like lipids (Subczynski et al., 1992) provides a robust per- meability barrier against O_2 entering the cell. This mechanism deserves more atten- tion in myelinated neurons.

15.5   SUMMARY

It has been recognized for many years that mitochondria, which consume most of the O_2 in the body, likely protect themselves and their host cells against oxidative damage by sustaining low O_2 levels. Values for O_2 concentrations maintained within mitochondria are likely variable, with estimates ranging from 3 to 30 µM O_2. If mito- chondria do lower O_2 levels to this extent, then it becomes clear that mitochondrial respiration plays a major role in protecting their vast network of polyunsaturated mitochondrial membranes against chemical oxidation. Among all organs in the body, the brain stands out in terms of its extraordinary respiratory activity. Indeed, the human brain may harness its extreme respiratory capacity for the purpose of protecting its DHA-enriched neurons and highly polyunsaturated mitochondria. Maintaining low O_2 levels might be a double-edged sword. For example, the net O_2 consumption by brain mitochondria might lower available oxygen levels in brain tissue uncomfortably close to the lethal threshold reached within minutes following events such as heart attacks or strokes that shut down a continuous supply of O_2 to the brain. From the perspective of a long brain span, mitochondria stand out in scavenging O_2, protecting themselves and DHA-enriched axon membranes and synaptosomes against O_2-mediated oxidative damage. Localized O_2 avoidance by myelination of axons is presented as a hypothetical mechanism to help explain the extraordinary stability of DHA in the human brain.

This concept is consistent with how sperm avoid O_2 damage to their DHA- enriched flagellar membranes during fertilization. We suggest that a modest shift of the ratios between glycolysis and respiration in axons might significantly lower O_2 levels aimed at protecting DHA membranes, with implications for how cells in the rest of the body protect their highly polyunsaturated mitochondrial membranes. O_2 avoidance in axons might require two adaptations. The first involves myelination as a physical barrier against oxygen, minimizing membrane peroxidation. The second part of the axon model involves rebalancing of ATP production by increasing the contribution of glycolysis to axon bioenergetics. The net effect is proposed to be a lowering of O_2 levels on both leaflets of the membrane for the purpose of protecting unsaturated phospholipids against peroxidation and oxidative damage, thus enabling brain span. According to this model, mitochondria practice a version of O_2 avoid- ance in which O_2 scavenging by respiration lowers O_2 levels and protects their highly polyunsaturated mitochondrial membranes.

REFERENCES

Berry, A. M., O. T. Harriott, R. A. Moreau, et al. 1993. Hopanoid lipids compose the *Frankia* vesicle envelope, presumptive barrier of oxygen diffusion to nitrogenase. *Proc. Natl. Acad. Sci. USA* 90:6091–94.

Chazotte, B., and C. R. Hackenbrock. 1988. The multicollisional, obstructed, long-range diffusional nature of mitochondrial electron transport. *J. Biol. Chem.* 263:14359–67.

Dalle-Donne, I., R. Rossi, R. Colombo, et al. 2006. Biomarkers of oxidative damage in human disease. *Clin. Chem.* 52:601–23.

DeMar Jr., J. C., K. Ma, J. M. Bell, et al. 2006. One generation of n-3 polyunsaturated fatty acid deprivation increases depression and aggression test scores in rats. *J. Lipid Res.* 47:172–80.

Erecińska, M., and I. A. Silver. 2001. Tissue oxygen tension and brain sensitivity to hypoxia. *Respir. Physiol.* 128:263–76.

Joshi, A. S., J. Zhou, V. M. Gohil, et al. 2009. Cellular functions of cardiolipin in yeast. *Biochim. Biophys. Acta* 1793:212–18.

Kubicki, M., R. McCarley, C. F. Westin, et al. 2007. A review of diffusion tensor imaging studies in schizophrenia. *J. Psychiatr. Res.* 41:15–30.

Leedo, Y., Z. Xun, V. Platt, et al. 2013. Distinct pools of non-glycolytic substrates differentiate brain regions and prime region-specific responses of mitochondria. *PLoS One* 8:e68831. doi: 10.1371/journal.pone.0068831.

Miki, K., W. Qu, E. H. Goulding, et al. 2004. Glyceraldehyde 3-phosphate dehydrogenase-S, a sperm-specific glycolytic enzyme, is required for sperm motility and male fertility. *Proc. Natl. Acad. Sci. USA* 101:16501–6.

Montine, T. J., J. F. Quinn, J. Kaye, et al. 2007. F(2)-isoprostanes as biomarkers of late-onset Alzheimer's disease. *J. Mol. Neurosci.* 33:114–19.

Montine, T. J., J. F. Quinn, D. Milatovic, et al. 2002. Peripheral F2-isoprostanes and F4-neuroprostanes are not increased in Alzheimer's disease. *Ann. Neurol.* 52:175–79.

Montine, T. J., J. F. Quinn, K. S. Montine, et al. 2005. Quantitative in vivo biomarkers of oxidative damage and their application to the diagnosis and management of Alzheimer's disease. *J. Alzheimers Dis.* 8:359–67.

Mukai, C., and M. Okuno. 2004. Glycolysis plays a major role for adenosine triphosphate supplementation in mouse sperm flagellar movement. *Biol. Reprod.* 71:540–47.

Murphy, M. P. 2009. How mitochondria produce reactive oxygen species. *Biochem. J.* 417:1–13.

Ndubuizu, O., and J. C. LaManna. 2007. Brain tissue oxygen concentration measurements. *Antioxid. Redox Signal.* 9:1207–19.

Phelps, M. E. 2000. Positron emission tomography provides molecular imaging of biological processes. *Proc. Natl. Acad. Sci. USA* 97:9226–33.

Poon, H. F., V. Calabrese, G. Scapagnini, et al. 2004. Free radicals and brain aging. *Clin. Geriatr. Med.* 20:329–59.

Skulachev, V. P. 1996. Role of uncoupled and non-coupled oxidations in maintenance of safely low levels of oxygen and its one-electron reductants. *Q. Rev. Biophys.* 29:169–202.

Subczynski, W. K., L. E. Hopwood, and J. S. Hyde. 1992. Is the mammalian cell plasma membrane a barrier to oxygen transport? *J. Gen. Physiol.* 100:69–87.

Turrens, J. F. 2003. Mitochondrial formation of reactive oxygen species. *J. Physiol.* 552:335–44.

Umhau, J. C., W. Zhou, R. E. Carson, et al. 2009. Imaging incorporation of circulating docosahexaenoic acid into the human brain using positron emission tomography. *J. Lipid Res.* 50:1259–68.

Young, R. W. 1967. The renewal of photoreceptor cell outer segments. *J. Cell Biol.* 33:61–72.

16 Uncoupling Proteins (UCPs) of Mitochondria Purposely Waste Energy to Prevent Membrane Damage

Mitochondria have evolved a sophisticated, energy-dependent feedback mechanism mediated by uncoupling proteins (UCPs) to prevent membrane peroxidation (Figure 16.1a). In 2000 Brand proposed the concept that purposeful proton uncoupling by mitochondria is essential for longevity (Brand, 2000), and recent data from his laboratory support this concept (Divakaruni and Brand, 2011; Jastroch et al., 2010). The purpose of this chapter is to develop an overview as to how purposeful uncoupling of proton gradients by uncoupling proteins at the level of the inner mitochondrial membrane is as a mechanism to prevent membrane peroxidation and enhance longevity.

The mechanism proposed by Brand (2000) involves the evolution of a family of UCPs embedded in the inner mitochondrial membrane, along with a sophisticated regulatory feedback loop that acts to prevent membrane peroxidation by minimizing production of reactive oxygen species (ROS) during mitochondrial respiration. Perhaps the biggest surprise is that continuous and purposeful proton uncoupling (energy uncoupling) is the price tag to accomplish this goal. However, the energy cost may be a bargain since longevity itself seems to be the reward. For a historical overview of the roles of uncoupling proteins in governing ROS production, see Mookerjee et al. (2010).

16.1 NATURE OF MITOCHONDRIAL ENERGY UCPS AND THEIR ACTIVATION BY PUFAS, HUFAS, AND FATTY ACID PEROXIDATION PRODUCTS

The x-ray structure of mitochondrial uncoupling proteins (five homologs in humans) is not yet available, but a nuclear magnetic resonance (NMR)-generated picture of a related mitochondrial protein (adenine nucleotide translocase (ANT)) has been created (Figure 16.1b) (Berardi et al., 2011). The structure of ANT is shown on the left of this figure, with UCP2 on the right. UCP2 is an integral membrane protein that is expressed ubiquitously, but at low levels in mammalian mitochondria. This

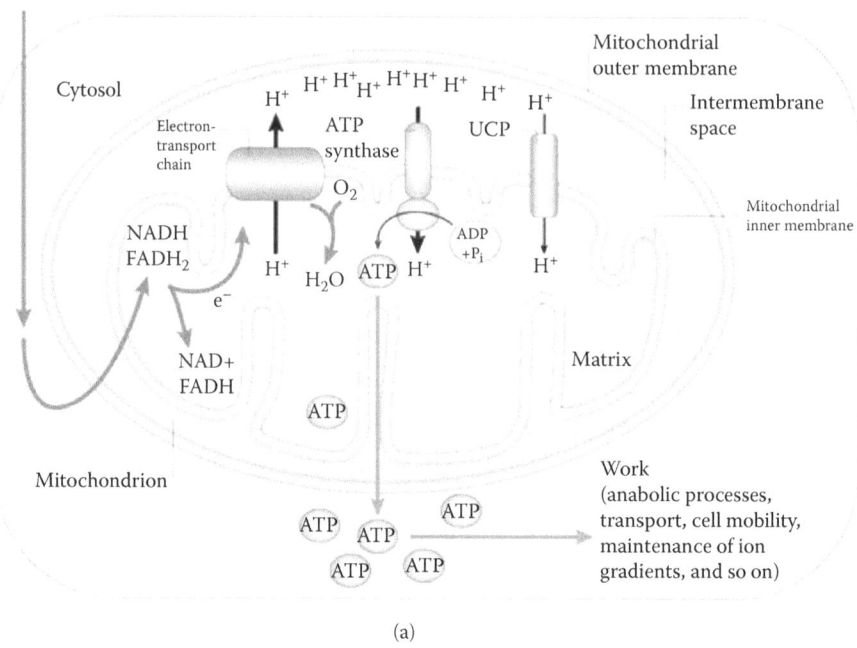

(a)

FIGURE 16.1 **See color insert.** UCPs 2–5 dissipate energy held in proton gradients and help protect the polyunsaturated fatty acid (PUFA)-enriched inner mitochondrial membrane against an oxidative chain reaction. (a) Diagram integrates mitochondrial bioenergetics and shows that UCPs act to uncouple the proton electrochemical gradients of mitochondria. See text for details. (From Krauss et al., *Nat. Rev. Mol. Cell. Biol.* 6:248–61, 2005. Copyright © 2005. Reprinted by permission from Macmillan Publishers Ltd.) *(continued)*

protein has six transmembrane domains with pairs of membrane-spanning segments of the protein being joined on the matrix side by loops. The embedded protein forms a pore, which in the case of ANT catalyzes ATP efflux out of the mitochondrion with influx of ADP needed as the phosphoryl acceptor for maintaining oxidative phosphorylation. The first member of the uncoupling protein family to be discovered was UCP1, expressed at levels of 10 to 15 percent of total membrane protein in mitochondria of brown adipose tissue. Energy stored as fat in brown adipose tissue is converted to heat by energy-uncoupled mitochondria to maintain or warm vital body organs or tissues. Following adaptation to the cold in mammals such as rats, UCP1 levels rise sharply in these specialized, heat-producing mitochondria. The requirement for such high levels of UCP1 to catalyze heat production is a clue that the mode of action of UCP1 might be rather sluggish compared to rates of biocatalysis by other enzymes. UCP2 is expressed in most cells, but at much lower levels than UCP1 in adipose tissue, suggesting that heat production is not the purpose of this uncoupling protein. This idea is consistent with the finding that, when compared on

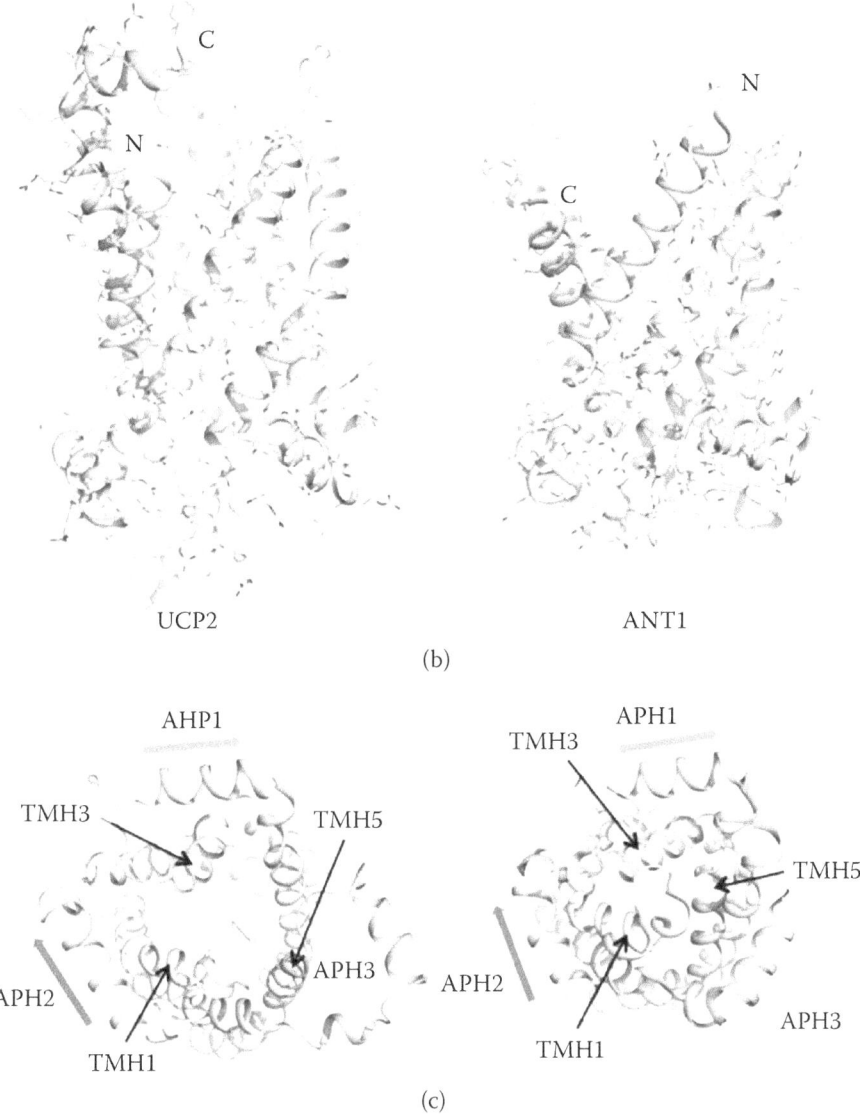

FIGURE 16.1 *(continued)*   See color insert. UCPs 2–5 dissipate energy held in proton gradients and help protect the polyunsaturated fatty acid (PUFA)-enriched inner mitochondrial membrane against an oxidative chain reaction. (b and c) Close structural similarity between UCP2 and ANT1 suggests UCPs might have evolved from ANT. Figures provide a side view (at top) and top view (at bottom) for UCP2 and ANT1. Note that ANT is one of the most abundant proteins in the inner membranes of mitochondria and has been shown to display unregulated proton uncoupling activity. (From Berardi et al., *Nature* 476:109–13, 2011. Copyright © 2011. Reprinted by permission from Macmillan Publishers Ltd.)

a molecule-by-molecule basis as studied in a planar membrane system, UCP1 and UCP2 have similar enzymatic powers and requirements for activation (Beck et al., 2007). It is known that the family of UCPs has some unusual mechanisms required for activation (Echtay et al., 2002).

It has taken several years to develop a model for the mode of action of UCPs, and there are still important unanswered questions and some controversy. Nevertheless, there is a consensus that UCPs have evolved a novel mode of action, including the requirement for long-chain fatty acids as activators (Klingenberg and Huang, 1999; Fedorenko et al., 2012) (Figure 16.2). Several laboratories that study the mechanism of action of fatty acids in UCPs have developed models to explain their data (Garlid et al., 2000; Goglia and Skulachev, 2003; Jaburek et al., 2004; Beck et al., 2007;

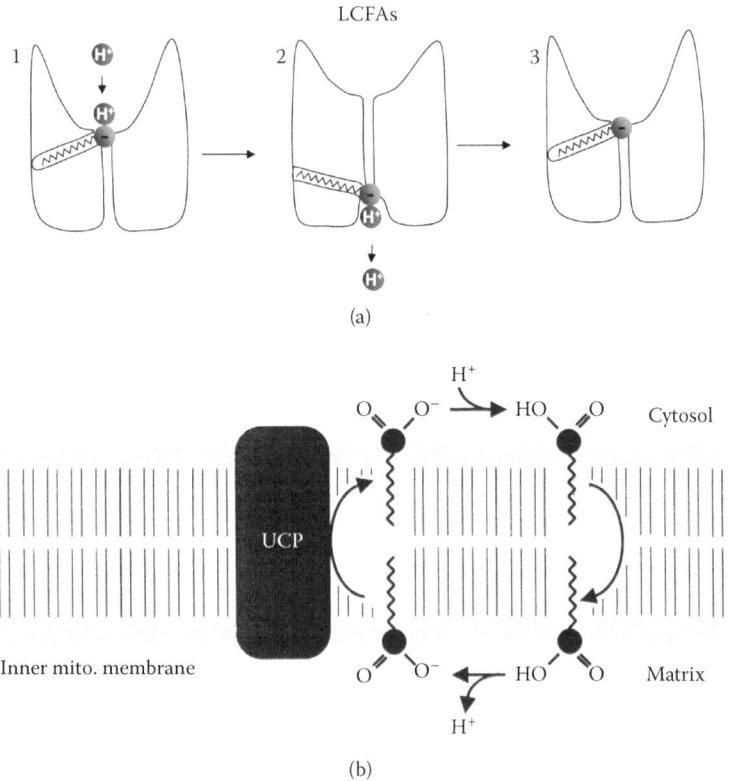

FIGURE 16.2 (a) Long-chain fatty acids (LCFAs) as coenzymes with UCPs mediating proton energy uncoupling in mitochondria. In this model (b) a membrane-bound uncoupling protein operates as a symporter that transports one LCFA and one H+ per transport cycle. The LCFA anion first binds to UCP on the cytosolic side of the mitochondrial membrane. H+ binding to UCP occurs only after the LCFA anion binds to UCP. The H+ and the LCFA are translocated by the UCP following a major conformational change, and H+ is released on the opposite side of the mitochondrial inner membrane (IMM). The LCFA stays associated with the UCP due to the hydrophobic interactions established by its long carbon tail. The LCFA anion then returns to initiate another H+ translocation cycle. (From Fedorenko et al., *Cell* 151:400–13, 2012. Copyright © 2012. With permission from Elsevier.)

Fedorenko et al., 2012). Previous researchers considered saturated fatty acids such as palmitic acid (16:0) to be the standard for measuring fatty acid specificity of activation of UCPs. Beck and coworkers (2007) used a well-defined system of planar lipids to study the roles of fatty acids as activators of UCPs. These researchers compared the role of fatty acids as mediators of proton conductance by UCP1 versus UCP2. One of the main conclusions is that compared on a molecule-to-molecule basis, rates of proton conductance by UCP1 and UCP2, which share about 60 percent sequence homology, are similar. Both uncoupling proteins require fatty acids for activity, and these data show that the highly unsaturated fatty acid (HUFA) 20:4 is the most active in the order arachidonic > retinoic > linoleic > eicosatrienoic > oleic > stearic. Beck and colleagues (2007) developed a provocative model based on membrane lipid dynamics to explain how fatty acids act as co-catalysts in shuttling protons across the inner mitochondrial membrane (i.e., from the matrix to the cellular cytoplasm).

Goglia and Skulachev (2003) originally suggested that UCPs might play a defensive role against oxidative stress by flipping peroxidized fatty acids from the inner mitochondrial membrane to the outer leaflet. In 2004, Jaburek and coworkers (2004) showed that linoleic acid (18:2) hydroperoxides (LAOOH), expected to be the major class of fatty acid peroxides generated during oxidative damage of human mitochondrial membranes, act as a potent activator of proton conductance by UCP2. Affinity (Km) for 18:2-hydroperoxide is three-fold greater than for linoleic acid (18:2), which was shown by Beck and coworkers (2007) to be almost double that of 18:1. Interestingly, the peroxide of 18:2 also mediates K^+ influx by UCP2, showing that the cation specificity of UCPs extends to other important cations. These findings by Jaburek and colleagues (2004) led these authors to propose an interesting model to explain activation of UCP by ROS. ROS generated by the respiratory chain and facilitated by Fenton chemistry are proposed to generate hydroperoxides and other peroxidation products. Fatty acid peroxides can be cleaved from the phospholipids by phospholipases, including PLA_2. Hydroperoxide fatty acid can recycle, being effluxed from the matrix via UCP2 in an anionic form, and carries a H^+ and creates uncoupling while returning back in its protonated form. Such uncoupling can attenuate mitochondrial superoxide production at the complex 1 site (Chapter 11). Consequently, a feedback down-regulation of ROS is achieved. The ability of fatty acid peroxides to induce UCP2-mediated H^+ uniport points to the essential role of superoxide reaction products such as hydroperoxyl radical, hydroxyl radical, or peroxynitrite initiating lipoperoxidation. The main point is by-products of membrane peroxidation appear to activate the UCP2-mediated uncoupling and promote the feedback down-regulation of reactive oxygen species during mitochondrial respiration.

16.2   A MOLECULAR MODEL LINKING MEMBRANE UNSATURATION WITH LONGEVITY

Among the five homologs of UCPs found in humans, UCP1 is implicated in thermogenesis and basal mitochondrial proton leakage. UCPs 2–5 are suggested to play crucial roles in mediating proton conductance linked to mitochondrial bioenergetics and modulation of production of reactive oxygen species by the electron transport chain. The importance of UCPs in aging and age-related diseases is drawing increasing attention (Brand, 2000; Wolkow and Iser, 2006; Mookerjee et al., 2010; Dietrich and Horvath, 2010; Manini, 2010; Rose et al., 2011; Mailloux and Harper, 2011; Arsenijevic et al., 2000; Vidal-Puig et al., 2000). We do not attempt here to review these papers, which are mentioned because these data illustrate how different segments of the complex feedback loop between membrane unsaturation, ROS, and longevity are being linked together through data derived from different laboratories. The combined reference lists, especially those in the most recent papers above, show the growing interest in this field and provide a snapshot of research on the relationships between mitochondrial bioenergetics and aging. Several recent milestones toward establishing the bioenergetic basis of aging are listed as follows:

- Complex 1 is a major source of ROS.
- Rate of ROS production by complex 1 is closely regulated.
- Rates of proton leakage via UCPs are modulated by ROS.
- Modulation of ROS levels in mitochondria is energy dependent and adds to overall energy stress.
- Net energy stress from all sources causes aging.

The discovery of UCPs and elucidation of their roles in governing ROS levels are consistent with the importance of protecting mitochondrial membranes. These data also show that humans and other organisms are prepared to pay the high bioenergetic price of this protection. Note that UCPs geared to protecting mitochondria generally use relatively small amounts of the total energy produced during respiration. However, this cost accrues over a lifetime, and thus represents a significant bioenergetic burden. Also note that energy production drops with aging due to accumulation of mutations in key electron transport components. If it is assumed that energy consumption to protect mitochondria is constantly high, then the percentage of energy produced compared to energy needed for protection will increase with aging. We propose that this ratio (energy produced:energy utilized for oxidative protection) not only increases with aging, but also is a gauge defining the health of mitochondria and predicts when massive programmed cellular death will occur. This concept is a mechanistic explanation for the peroxidation theory of aging in which unsaturation levels of mitochondrial membranes correctly predict life span of mammals. It is interesting to speculate that if this ratio could be determined for an individual, it would help monitor the course of aging.

16.3 SUMMARY

The evolution of mitochondrial uncoupling proteins (UCPs), which uncouple proton gradients in mitochondria, forming heat, would seem to be an exercise in bioenergetic futility. A closer look provides a new perspective on the energy cost of mechanisms that provide a long-term insurance policy against oxidative stress triggering apoptosis. The main point is that protecting membranes not only saves a great deal of energy, but also ensures that a catastrophic chain reaction such as occurs with paraquat poisoning (Chapter 2) is never triggered. It is now clear that membrane protection consumes significant amounts of energy balanced against long-term benefits. According to the model proposed by Brand (2000), UCPs are an integral part of a sophisticated energy uncoupling feedback loop that down-regulates production of ROS by mitochondria. Implicit in this model is the assumption that down-regulation of ROS production protects cellular constituents against excessive oxidative damage. Membranes have long been considered to be targets of ROS damage but are not often highlighted as we describe here. Indeed, at the time that Brand (2000) proposed his model featuring UCPs, DNA, not membranes, was considered to be the major target of ROS damage, followed by proteins and membranes in order of importance. Now, more than a decade after Brand's model for UCPs appeared, one prediction seems riveting to us. This point involves energy uncoupling as a mechanism of providing long-term protection of mitochondrial membranes against oxidative damage. The overall bioenergetic cost seems high unless balanced against some extraordinary gain—a benefit that we suggest includes protecting the vast network of mitochondrial membranes, allowing mitochondria to function longer and more efficiently.

REFERENCES

Arsenijevic, D., H. Onuma, C. Pecqueur, et al. 2000. Disruption of the uncoupling protein-2 gene in mice reveals a role in immunity and reactive oxygen species production. *Nat. Genet.* 26:435–39.

Beck, V., M. Jabůrek, T. Demina, et al. 2007. Polyunsaturated fatty acids activate human uncoupling proteins 1 and 2 in planar lipid bilayers. *FASEB J.* 21:1137–44.

Berardi, M. J., W. M. Shih, S. C. Harrison, et al. 2011. Mitochondrial uncoupling protein 2 structure determined by NMR molecular fragment searching. *Nature* 476:109–13.

Brand, M. D. 2000. Uncoupling to survive? The role of mitochondrial inefficiency in ageing. *Exp. Gerontol.* 35:811–20.

Dietrich, M. O., and T. L. Horvath. 2010. The role of mitochondrial uncoupling proteins in lifespan. *Pflugers Arch.* 459:269–75.

Divakaruni, A. S., and M. D. Brand. 2011. The regulation and physiology of mitochondrial proton leak. *Physiology (Bethesda)* 26:192–205.

Echtay, K. S., D. Roussel, J. St.-Pierre, et al. 2002. Superoxide activates mitochondrial uncoupling proteins. *Nature* 415:96–99.

Fedorenko, A., P. V. Lishko, and Y. Kirichok. 2012. Mechanism of fatty-acid-dependent UCP1 uncoupling in brown fat mitochondria. *Cell* 151:400–13.

Garlid, K. D., M. Jabůrek, P. Ježek, et al. 2000. How do uncoupling proteins uncouple? *Biochim. Biophys. Acta* 1459:383–89.

Goglia, F., and V. P. Skulachev. 2003. A function for novel uncoupling proteins: antioxidant defense of mitochondrial matrix by translocating fatty acid peroxides from the inner to the outer membrane leaflet. *FASEB J.* 17:1585–91.

Jaburek, M., S. Miyamoto, P. Di Mascio, et al. 2004. Hydroperoxy fatty acid cycling mediated by mitochondrial uncoupling protein UCP2. *J. Biol. Chem.* 279:53097–102.

Jastroch, M., A. S. Divakaruni, S. Mookerjee, et al. 2010. Mitochondrial proton and electron leaks. *Essays Biochem.* 47:53–67.

Klingenberg, M., and S. G. Huang. 1999. Structure and function of the uncoupling protein from brown adipose tissue. *Biochim. Biophys. Acta* 1415:271–96.

Krauss, S., C. Y. Zhang, and B. B. Lowell. 2005. The mitochondrial uncoupling-protein homologues. *Nat. Rev. Mol. Cell. Biol.* 6:248–61.

Mailloux, R. J., and M. E. Harper. 2011. Uncoupling proteins and the control of mitochondrial reactive oxygen species production. *Free Radic. Biol. Med.* 51:1106–15.

Manini, T. M. 2010. Energy expenditure and aging. *Ageing Res. Rev.* 9:1–11.

Mookerjee, S. A., A. S. Divakaruni, M. Jastroch, et al. 2010. Mitochondrial uncoupling and lifespan. *Mech. Ageing Dev.* 131:463–72.

Rose, G., P. Crocco, F. De Rango, et al. 2011. Further support to the uncoupling-to-survive theory: the genetic variation of human UCP genes is associated with longevity. *PLoS One* 6:e29650.

Vidal-Puig, A. J., D. Grujic, C. Y. Zhang, et al. 2000. Energy metabolism in uncoupling protein 3 gene knockout mice. *J. Biol. Chem.* 275:16258–66.

Wolkow, C. A., and W. B. Iser. 2006. Uncoupling protein homologs may provide a link between mitochondria, metabolism and lifespan. *Ageing Res. Rev.* 5:196–208.

17 Mitochondrial Fission Protects against Oxidative Stress by Minting a Continuous Supply of Cardiolipin and Other Polyunsaturated Phospholipids

Mitochondria in liver cells of a centenarian are estimated to have divided by fission (Figure 17.1) as many as 3600 times. It is clear that division acts to rejuvenate worn-out parts of mitochondria that might otherwise decrease both energy output and efficiency. In Chapter 12 we discussed how specialized membranes of rhodopsin disks senesce on a timescale of days, and how tails of sperm cells, when exposed to ambient O_2 in the test tube, are degraded on a scale of hours. Data on mitochondria show that the functional life span of a single liver mitochondrion, measured as the time between divisions, is as short as ten days—roughly equivalent to the turnover time of a rhodopsin membrane disk in a rod cell of the eye. We suggest that the human inner mitochondrial membrane is another example of such a specialized membrane based on its extraordinarily high percentage of polyunsaturated fatty acids, the presence of cardiolipin molecular species with four 18:2 chains, and its rapid turnover. We also propose that the relatively rapid rates of mitochondrial division evolved in part as a mechanism for rejuvenating the vast and oxidatively vulnerable inner membrane, thus avoiding effects from accumulation of oxidatively damaged cardiolipin and other phospholipids. Even though the energy cost of fission is high, the benefits are great and include the continuous production of virgin membrane surface, which avoids an oxidative chain reaction. The benefits and risks of mitochondrial fission are so powerful as to impact longevity.

17.1 DISCOVERY OF MAMs

Mitochondrial membranes age rapidly and require renewal just like rhodopsin disks of rod cells (Chapter 12). The caveats are that in mitochondria, energy homeostasis,

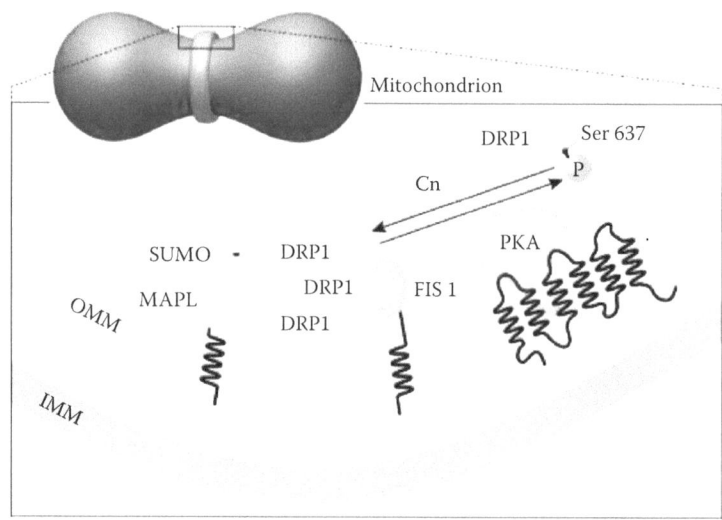

FIGURE 17.1   See color insert. Diagram showing mechanism of formation of a fission "collar" generated by dynamic trafficking of dynamin-like protein DRP1 to the site of fission. A putative network regulating DRP1 accumulation and assembly is shown, in which DRP1 translocation is controlled by calcineurin-mediated dephosphorylation of Ser 637. Mitochondrial PKA (phosphokinase) then rephosphorylates the same site, pushing DRP1 away from the organelle. MAPL-mediated SUMOylation stabilizes DRP1 on mitochondria and might prevent its re-translocation to the cytoplasm. FIS1, fission 1; IMM, inner mitochondrial membrane; MAPL, mitochondrial-anchored protein ligase; OMM, outer mitochondrial membrane; PKA, protein kinase A; SUMO, small ubiquitin-like modifier. [Reprinted by permission from Macmillan Publishers Ltd: *EMBO Rep.* 10:694-6. Scorrano, L. and D. Liu. The SUMO arena goes mitochondrial with MAPL, © 2009.]

not vision, is at stake. We suggest that renewal of mitochondrial membranes in humans is tailored to long-term bioenergetic needs. Unlike bacteria, which usually synthesize their own membrane phospholipids, mitochondria depend on transport of phospholipids between membranes of the endoplasmic reticulum (ER) and mitochondria (Osman et al., 2011). These researchers show that phospholipid transfer from ER ⇔ mitochondria occurs at specialized regions of the ER that are tightly associated with mitochondria and are named MAMs, standing for mitochondria-associated membranes.

Yeast has turned out to be an important research tool in elucidating the mechanism of tethering between the ER and mitochondria (Figure 17.2). In yeast the tethering complex is composed of an integral ER glycoprotein (Mmm1) and three mitochondria-associated proteins (Mdm34, Mdm10, and Mdm12). This complex promotes and stabilizes interactions between the two membranes.

An intimate contact between ER and mitochondria opens up the possibility of novel mechanisms of phospholipid transfer, perhaps bypassing the need for lipid trafficking vesicles. However, relatively little is known about how newly imported phospholipids or lipid precursors are transported within mitochondria. The current

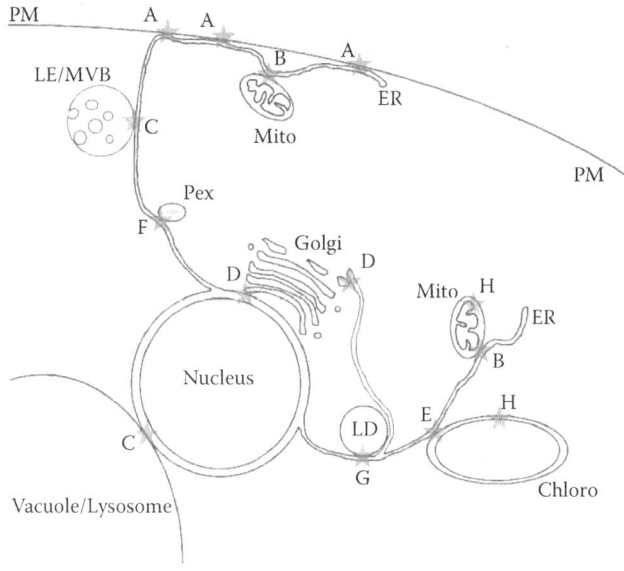

FIGURE 17.2 This figure summarizes how new membrane surface is generated during mitochondrial fission with emphasis on location of membrane contact sites. A. Endoplasmic reticulum (ER) - plasma membrane (PM) B. ER - Mitochondria (Mito) C. ER - Late endosome (LE) D. ER – Golgi complex E. ER and chloroplast (chloro) F. ER – peroxisome (Pex) G. ER and lipid droplets (LD) H. Contact sites between the inner and outer membranes of mitochondria and chloroplasts. [Reprinted from *Curr. Opin. Cell Biol.*, 23, Toulmay, A. and W. A. Prinz, Lipid transfer and signaling at organelle contact sites: the tip of the iceberg, pages 458-63, © 2011, with permission from Elsevier.]

working model is that mechanisms have evolved for phospholipid flip-flop from one leaflet to the other. Recent data suggest that laying down a new mitochondrial membrane surface, as well as proofreading and repair of defects, is a highly sophisticated process of sufficient importance to impact the long-term health of cells and organisms.

17.2 MAJOR MOLECULAR SPECIES OF MITOCHONDRIAL PHOSPHOLIPIDS AND THEIR BIOSYNTHESIS

The major molecular species of mitochondrial phospholipids and their pathways of biosynthesis are understood in detail (Figure 17.3) (Osman et al., 2011; Lebiedzinska et al., 2009). For example, the most abundant phospholipids are phosphatidylcholine (PC) and phosphatidylethanolamine (PE), comprising approximately 40 and 30 percent of total mitochondrial phospholipids, respectively. Cardiolipin (CL) and phosphatidylinositol (PI) account for 10 to 15 percent of phospholipids. Phosphatidic acid (PA) and phosphatidylserine (PS) comprise about 5 percent of total mitochondrial phospholipids.

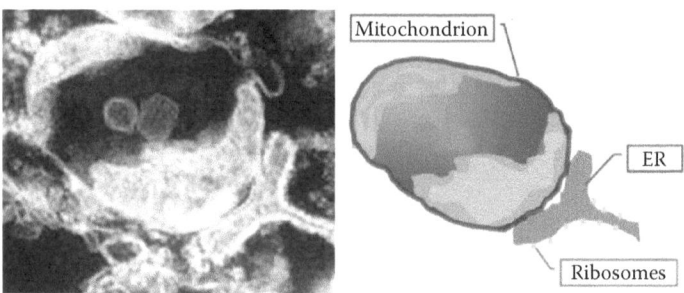

FIGURE 17.3 Association of the endoplasmic reticulum (ER) with a mitochondrion allows lipid exchange. Electron micrograph made by Mariusz R. Wieckowski (left panel) with traced profiles of mitochondrion and ER (right panel). [Reprinted from Int. *J. Biochem. Cell Biol.*, 41, Lebiedzinska, M., G. Szabadkai, A. W. Jones et al, Interactions between the endoplasmic reticulum, mitochondria, plasma membrane and other subcellular organelles, pages 1805-16, © 2009, with permission from Elsevier.]

Phospholipid biosynthesis carried out by mitochondria and their host cell illustrates a classic example of a biochemical partnership between a cell and its permanent endosymbiotic mitochondria (Figure 17.4). Mitochondria have retained the capacity to synthesize CL, PE, PG, and PA, but depend on cellular synthesis of PI, PC, and PS. Mitochondria produce an excess of PE, enough to supply themselves as well as contributing to cellular needs for this important membrane building block.

Cardiolipin, of special interest here, is synthesized by a multienzyme pathway in the mitochondrial inner membrane. The precursor cytidine diphosphate–diacylglycerol (CDP-DAG) initiates the pathway. CDP-DAG is cleaved yielding phosphatidylglycerol phosphate (PGP), which is next dephosphorylated to phosphatidylglycerol (PG). PG is condensed yielding immature CL, which through the action of the enzyme taffazin generates mature human CL—$(18:2)_4$-CL. At least two enzymes of the CL pathway in yeast are localized on the matrix side of the inner membrane, consistent with proposed initiation of CL synthesis on the inner leaflet of the inner membrane (Joshi et al., 2009). Whereas CL synthase generates immature human CL on the matrix side of the membrane, mature CL, which involves acyl chain remodeling, seems to be formed on the outer leaflet of the inner membrane. CL remodeling requires the sequential action of a phospholipase and a transacylation reaction catalyzed by taffazin. As discussed in Chapter 8, mutations in taffazin cause cardiomyopathy and Barth syndrome in humans.

Whether or not mitochondria can carry out the complete synthesis of CL starting with PA is not clear, but it seems likely. PA synthesis is known to occur in the ER and might be transported to mitochondria for CL synthesis. On the other hand, PA might be generated within mitochondria, raising the possibility that dual sources of phospholipid precursors for CL formation are available.

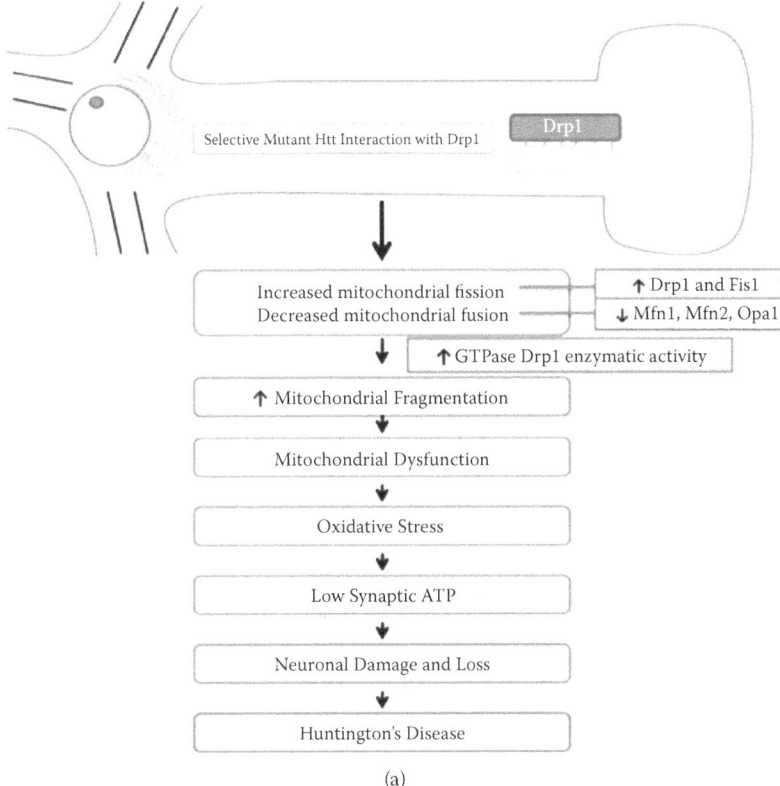

Selective Mutant Htt Interaction with Drp1 | Drp1

Increased mitochondrial fission ———— ↑ Drp1 and Fis1
Decreased mitochondrial fusion ———— ↓ Mfn1, Mfn2, Opa1

↑ GTPase Drp1 enzymatic activity

↑ Mitochondrial Fragmentation

Mitochondrial Dysfunction

Oxidative Stress

Low Synaptic ATP

Neuronal Damage and Loss

Huntington's Disease

(a)

FIGURE 17.4 Mutant huntingtin protein may act by disrupting the delicate balance between fission and fusion. a) Highly polyunsaturated mitochondrial membranes might contribute to oxidative and energy stresses facilitating Huntington's disease and other neurodegenerative diseases. *(continued)*

17.3 EXCESSIVE MITOCHONDRIAL FISSION MAY GENERATE TOXIC MITOCHONDRIA

It is now clear that a delicate balancing act occurs between mitochondrial fission or division and fusion. Mitochondrial fusion has already been introduced in Chapter 9 as an antiaging mechanism. However, there is increasing evidence discussed below that cells, especially neurons, must constantly adjust the ratio of fission to fusion during aging. For purposes of discussion here, mitochondria present in, say, a neuron can be divided into three classes, as follows:

- Normal mitochondria fulfilling their role as ATP machines
- Mitochondria destined to be or being rejuvenated by fission
- Unhealthy, unwanted, or toxic mitochondria marked for mitophagy, as discussed in Chapter 18

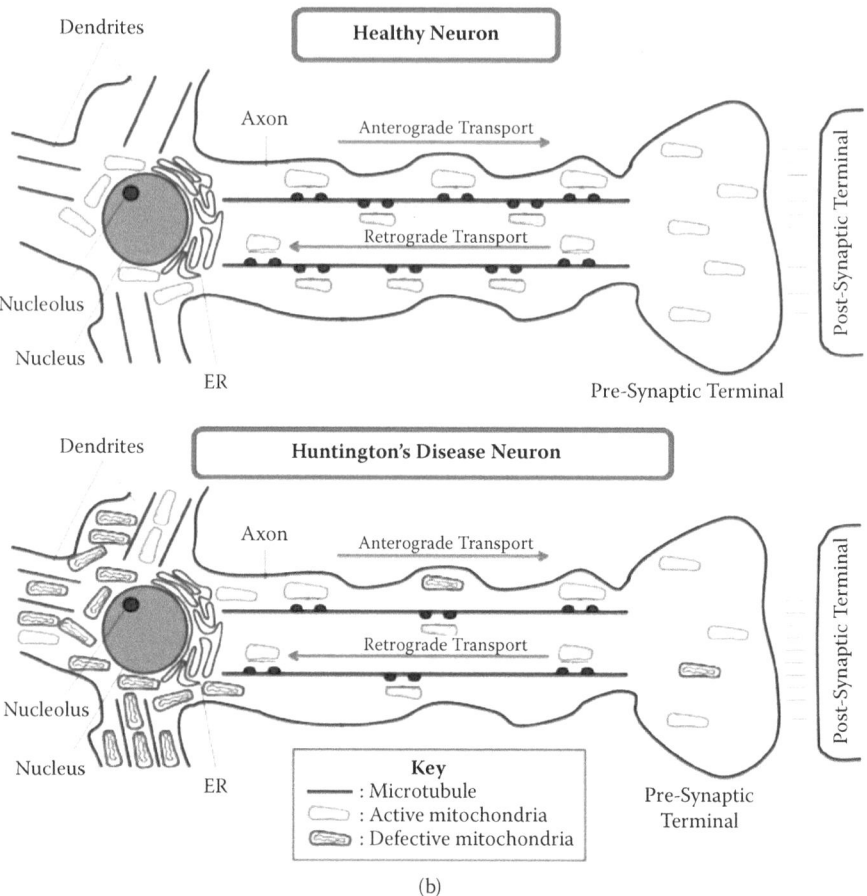

FIGURE 17.4 *(continued)* Mutant huntingtin protein may act by disrupting the delicate balance between fission and fusion. (b) Excessive fragmentation and abnormal distribution may enable pathogenicity of toxic mitochondria. (From Hemachandra Reddy and Shirendeb, *Biochim. Biophys. Acta* 1822:101–10, 2012. Copyright © 2012. Reprinted with permission from Elsevier.)

As discussed in Chapter 9, mutational analysis in mice shows that excessive mitochondrial fission in mutants blocked for fusion results in miniaturization of mitochondria with potential pathological consequences. For example, the number of fully functional mtDNA copies in miniature mitochondria can be lowered dramatically, which means that pathologies associated with mutated classes of mtDNA can become unmasked. The main points are that such mini-mitochondria may be beyond repair by fission, and indeed might have reached a stage in which they become toxic. Toxic mitochondria are defined in more detail in Chapter 18. But a key point made here is that the highly beneficial role for polyunsaturated membranes seen in normal mitochondria might be switched in mini-mitochondria into a harmful or toxic state.

Prior to the point at which dysfunctional mitochondria are dismantled by mitophagy, their dysfunctional electron transport chains can produce harmful levels of ROS.

There is increasing evidence that the balance between fission and fusion can be tipped during neurodegeneration when toxic proteins/peptides, such as in the case of mutant huntingtin (mHtt), bind to fission machinery (Johri et al., 2011). Mutant huntingtin has been found to bind to the mitochondrial fission GTPase dynamin-related protein-1 and increase its enzymatic activity (Song et al., 2011). Hemachandra Reddy and Shirendeb (2012) review this important finding and propose a unified concept of neurodegeneration based on the Huntington's disease model (Figure 17.4a,b). Figure 17.4a,b shows how mHtt interacts with the mitochondrial fission protein dynamin-related protein-1 (Drp1) and enhances GTPase Drp1 enzymatic activity. This effect causes excessive mitochondrial fragmentation and abnormal distribution, leading to defective axonal transport of mitochondria and eventually synaptic degeneration. The molecular pathology is not understood in detail, but both oxidative and energy stresses are implicated (Figure 17.4a). It is interesting to speculate that highly polyunsaturated membranes of dysfunctional mitochondria contribute to both oxidative stress and energy stress, another example of the benefits versus risks of membrane unsaturation. The discovery of the biochemical roles of huntingtin (Song et al., 2011) not only is an important milestone in understanding Huntington's disease (HD), but also has major implications for other neurodegenerative diseases (Manczak and Reddy, 2012; Calkins et al., 2011) and aging (Kowald and Kirkwood, 2011), the latter discussed next.

Figge and colleagues (2012) used a modeling approach to explain two interesting questions in the field of aging. Why are fission and fusion rates reduced during aging? Why does loss of a mitochondrial fission factor extend life span in fungi? These researchers propose the mitochondrial infectious damage adaptation (MIDA) model to answer these questions. According to this model, deceleration of fission-fusion cycles reflects a systemic adaptation increasing life span. This model predicts that preexisting molecular damage to mitochondria may be propagated and enhanced across the mitochondrial population by content mixing. In other words, with aging, dysfunctionality donated by one mitochondrion can spread like an infection. However, the benefits to the cell of fusion seem to outweigh this risk, as discussed in Chapter 9. The MIDA model is supported by data described above in which deceleration of fission-fusion may decrease neuronal death during Huntington's disease. Together, the data generated from the model and the data from HD suggest that rates of fission-fusion must be delicately regulated and require continuous rebalancing during aging to capture benefits. This helps explain why fission-fusion declines with aging and raises the possibility of an age-mediated regulatory cascade. A similar scenario might explain why a knockout mutation of a fission factor enhances the life span of fungi.

17.4 SUMMARY

Several important advances have recently been made in understanding the molecular biology of mitochondrial fission and its impact on aging and age-dependent diseases.

These data provide the clearest picture yet of how human cells use fission to continually generate virgin mitochondrial membranes essential for energy production. We consider fission to be at the heart of membrane protection against oxidative damage. Biochemistry of how the inner mitochondrial membrane is synthesized, including trafficking of fatty acids to specific membranes, is now coming to light. One area receiving increasing attention concerns the synthesis, repair, and roles of cardiolipin as a trigger for age-dependent diseases and aging. The discovery that Huntington's disease is an energy disease caused by an imbalance between fission and fusion has opened up new targets for managing HD and neurodegeneration with implications for aging. Perhaps the biggest surprise is that the rates of fission versus fusion act as a tipping point, switching from a benefit to a lethal risk. All of this exciting new data emerging from the field of mitochondrial fission-fusion seems consistent with the membrane perspective of aging described in this book.

REFERENCES

Calkins, M. J., M. Manczak, P. Mao, et al. 2011. Impaired mitochondrial biogenesis, defective axonal transport of mitochondria, abnormal mitochondrial dynamics and synaptic degeneration in a mouse model of Alzheimer's disease. *Hum. Mol. Genet.* 20:4515–29.

Figge, M. T., A. S. Reichert, M. Meyer-Hermann, et al. 2012. Deceleration of fusion-fission cycles improves mitochondrial quality control during aging. *PLoS Comput. Biol.* 8:e1002576. doi: 10.1371/journal.pcbi.1002576.

Hemachandra Reddy, P., and U. P. Shirendeb. 2012. Mutant huntingtin, abnormal mitochondrial dynamics, defective axonal transport of mitochondria, and selective synaptic degeneration in Huntington's disease. *Biochim. Biophys. Acta* 1822:101–10.

Johri, A., R. K. Chaturvedi, and M. F. Beal. 2011. Hugging tight in Huntington's. *Nat. Med.* 17:245–46.

Joshi, A. S., J. Zhou, V. M. Gohil, et al. 2009. Cellular functions of cardiolipin in yeast. *Biochim. Biophys. Acta* 1793:212–18.

Kowald, A., and T. B. Kirkwood. 2011. Evolution of the mitochondrial fusion-fission cycle and its role in aging. *Proc. Natl. Acad. Sci. USA* 108:10237–42.

Lebiedzinska, M., G. Szabadkai, A. W. Jones, et al. 2009. Interactions between the endoplasmic reticulum, mitochondria, plasma membrane and other subcellular organelles. *Int. J. Biochem. Cell Biol.* 41:1805–16.

Manczak, M., and P. H. Reddy. 2012. Abnormal interaction of VDAC1 with amyloid beta and phosphorylated tau causes mitochondrial dysfunction in Alzheimer's disease. *Hum. Mol. Genet.* 21:5131–46.

Osman, C., D. R. Voelker, and T. Langer. 2011. Making heads or tails of phospholipids in mitochondria. *J. Cell Biol.* 192:7–16.

Scorrano, L., and D. Liu. 2009. The SUMO arena goes mitochondrial with MAPL. *EMBO Rep.* 10:694–96.

Song, W., J. Chen, A. Petrilli, et al. 2011. Mutant huntingtin binds the mitochondrial fission GTPase dynamin-related protein-1 and increases its enzymatic activity. *Nat. Med.* 17:377–82.

Toulmay, A., and W. A. Prinz. 2011. Lipid transfer and signaling at organelle contact sites: the tip of the iceberg. *Curr. Opin. Cell Biol.* 23:458–63.

18 Mitophagy Eliminates Toxic Mitochondria

Recent data suggest that defective, unneeded, and senescing mitochondria generate toxic by-products, especially reactive oxygen species (ROS), threatening themselves, neighboring mitochondria, and the host cell. Toxic mitochondria arise continuously with aging and are eliminated by the process of mitophagy. Mitochondria marked for degradation are fenced off by encirclement with a membranous structure that fuses with a lysozyme where digestion proceeds. A detailed description of the biochemistry of mitophagy is beyond the scope of this book. Our interest focuses on the physiological benefits of mitophagy, especially in degrading toxic mitochondria and preventing oxidative and energy stress. Obviously, mitophagy rids the cell of severely oxidatively damaged membranes, in essence eliminating a major source of ROS.

18.1 BRIEF DESCRIPTION OF MITOPHAGY

The purpose of mitophagy is to eliminate unwanted, defective, or toxic mitochondria (Yorimitsu and Klionsky, 2005; Kim et al., 2007; Youle and Narendra, 2011; Green et al., 2011; Galluzzi et al., 2012; Levine et al., 2011). Mitophagy is considered to be a specialized form of autophagy in which cells degrade not only their mitochondria but also other organelles or even invading pathogens (Figure 18.1). Thus, autophagy removes worn-out or toxic elements from within the cell. Mitophagy clears mitochondria during maturation of red blood cells (Schweers et al., 2007), forcing these cells to obtain their energy by glycolysis. Data from analysis of knockout mutants of autophagy in mice illuminate the essential roles of this sophisticated and universal process. The journal *Autophagy* serves this growing field of science where increasing data show that autophagy and mitophagy are linked to aging, neurodegenerative diseases, and cancer (Green et al., 2011).

During mitophagy a damaged or superfluous mitochondrion is "tagged" and enclosed by a novel membranous structure called a phagophore, which elongates to form a double-membrane vesicle called an autophagosome. Autophagosomes fuse with lysosomes to form autolysosomes, where a battery of potent lysosomal hydrolases degrade the entrapped mitochondrion (Figure 18.1).

18.2 MITOPHAGY IN YEAST: A MECHANISM TO DECREASE OXIDATIVE STRESS

Just as humans use mitophagy to eliminate mitochondria during development of red blood cells, yeast cells subjected to nitrogen stress degrade an excess of mitochondria,

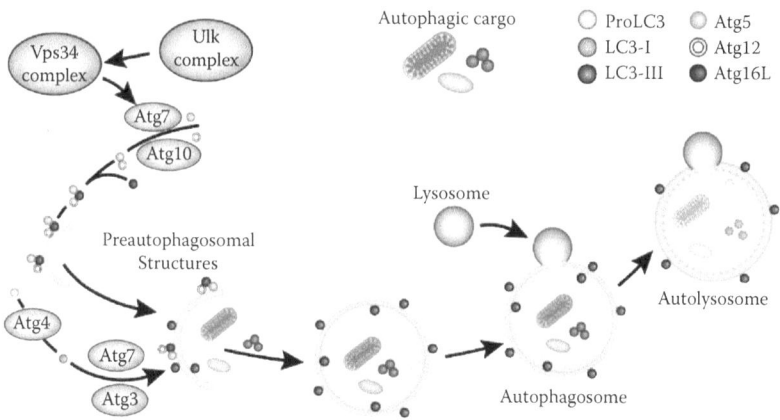

FIGURE 18.1 Mitophagy, a specific form of autophagy, degrades toxic mitochondria. This diagram shows how mammalian cells degrade mitochondria and other defective cellular components. In response to specific signals the ULK complex (consisting of ULK1/2-Atg13-FIP200-Atg101) initiates autophagy. The formation of the double-layered membrane encircling defective mitochondria within the cytosol requires the action of the Vps34 complex (Vps34, Vps15, Beclin1). The class III P13K activity of the complex is required for phagophore formation. Next, two conjugation systems add Atg12-Atg5-Atg16L complex and LC3-II to the encircling membrane. This elongating membrane grows to enwrap a portion of the cytosol, forming an autophagosome. In the final step of the process, lysosomes fuse with the autophagosome, releasing a potent battery of lysosomal hydrolases into the interior, resulting in degradation of mitochondria and other defective molecules and structures. (From Fleming et al., *Nat. Chem. Biol.* 7:9–17, 2011. Copyright © 2011. Reprinted by permission from Macmillan Publishers Ltd.)

which become unneeded as growth rate slows. One purpose is to release nitrogen held in mitochondrial proteins for synthesis of new proteins needed for maintenance by these nitrogen-starved cells. Recently new physiological roles of mitophagy in yeast have been reported, as follows (Suzuki et al., 2011; Kurihara et al., 2012):

- Down-regulation of ROS levels
- Decreased oxidative damage
- Decreased mutation rates of mitochondrial DNA

These roles were elucidated using knockout mutations of genes essential for mitophagy (Yorimitsu and Klionsky, 2005). Kurihara and colleagues (2012) propose that in the absence of mitophagy, defective mitochondria persist and become toxic. These authors propose that dysfunctional mitochondria produce dangerous levels of ROS, which damage mitochondrial membranes and other constituents, causing a vicious cycle of cellular destruction. Because of the complexities of the regulation and mechanisms of mitophagy, where many steps remain unknown, it is not yet possible to decipher a comprehensive molecular biology of mitophagy. However, data are now emerging that promise to answer some fundamental questions not only of mitophagy, but also of aging itself. For example, the field of aging remains plagued by a famous "chicken or egg" riddle regarding which comes first, oxidative stress or energy stress? Clearly, yeast cells formerly growing with an excess of both respiratory energy substrates and nitrogen sources, when subjected to a nitrogen-deficient medium, instantly face a potentially dangerous situation. According to the regulatory mechanism of complex 1 described by Pryde and Hirst (2011) the down-shift from high nitrogen to deficient nitrogen for growth is predicted to result in a burst of ROS (see Chapter 11). The rationale is that when nitrogen becomes limiting, cellular growth will slow significantly, causing the demand for ATP needed for growth to plummet and the level of NADH to rise. According to their model, these conditions are expected to increase ROS production mediated through the high-energy site of complex 1. Thus, the Pryde-Hirst model explains how nitrogen starvation and oxidative stress are linked, generating a ROS signal triggering the onset of mitophagy. This regulatory mechanism involving degradation of excess mitochondria in order to avoid ROS-mediated oxidative stress seems crude at first glance, until one considers possible benefits. As mentioned above, the first benefit is the availability of nitrogen from recycled proteins needed for protein synthesis, and the second benefit is a drop in levels of ROS, which might otherwise kill the cell.

An important question raised by studies of mitophagy in yeast concerns which potential cellular targets are most vulnerable to a burst of ROS caused by nitrogen stress. The three likely targets of ROS damage in yeast cells are membranes, proteins, and DNA. As discussed in Chapter 20, yeast membranes, being monounsaturated, represent the most stable class of unsaturated membranes in nature. This stability against peroxidation suggests that membranes of yeast are not the primary targets for ROS damage (see Chapter 20). However, yeast mitochondrial membranes often are enriched with CL with four successive monounsaturated chains (16:1 and 18:1), a structure that may be more sensitive to ROS damage. Enrichment of yeast membranes with as little as 5 percent of total phospholipids in the form of 18:2 greatly increases sensitivity of yeast cells to oxidative death (see Chapter 20).

18.3   ROLES OF MITOPHAGY IN MICE

Recent data suggest that mice, like yeast, may eliminate toxic mitochondria by mitophagy (Douarre et al., 2012). The term *toxic mitochondria* as applied to mammals requires some clarification and likely involves at least two classes of mitochondria. Damaged or dysfunctional mitochondria are the simplest class to visualize. A second class involves unwanted mitochondria similar to the case of nitrogen-starved yeast cells discussed above. The levels of mitochondria needed in a given tissue or cell such as liver can vary dependent on the workload or energy demand. Since the demand for energy can vary according to circumstances, a feedback loop similar to that in yeast might have evolved to prune an excess of mitochondria when they are not needed. Such unwanted mitochondria are not immediately defective in a conventional sense, but in excess they can become communally toxic. That is, too many mitochondria taking part in respiration can create a toxic environment by a mechanism similar to that described for yeast. Yeast solves this problem by pruning the total numbers of mitochondria, and mice do the same. Thus, a healthy, but unwanted mammalian mitochondrion might be tagged for elimination by mitophagy. Douarre and colleagues (2012) suggest that too many otherwise healthy mitochondria in a mouse cell might generate toxic levels of ROS. Obviously, mitophagy eliminates this threat. Physiologically or genetically damaged mitochondria that are beyond repair can also be classified as toxic, especially if ROS levels become elevated. The regulatory cross talk between ROS and mitophagy determining which mitochondria are to be dismantled is summarized in Figure 18.2.

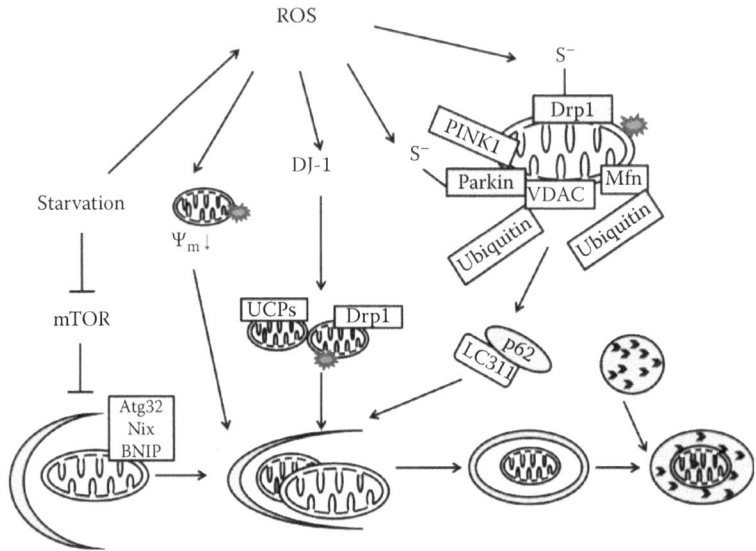

FIGURE 18.2 See color insert. Oxidative stress can activate mitophagy. An elaborate global signaling network allows mammalian cells to tag and eliminate defective or unwanted mitochondria by the process of mitophagy. General signals include oxidative stress and starvation; specific signals include signaling proteins or modification of mitochondrial proteins. In yeast, Atg32 is specific for mitophagy and acts by targeting mitochondria to the autophagosome. In mammalian cells, Nix is involved in mitochondrial clearance during maturation of erythrocytes. During energy stress, when ATP levels drop, AMPK is activated and phosphorylates ULK1 and ULK2 (both Atg1 homologs) in turn activate mitophagy and general autophagy. Parkinson's disease genes encoding a-synuclein, parkin, PINK1, and DJ-1 are all involved in mitophagy. A decrease in mitochondrial membrane potential (ψ_m) can be induced by ROS and by targeting a-synuclein to the mitochondria. A depression in mitochondrial membrane potential serves as a signal for mitophagy. In addition to a decrease in membrane potential, mitochondrial fission is another signal for mitophagy. PINK1 facilitates parkin targeting to the mitochondria and ubiquitinates the mitochondrial outer membrane VDAC. Ubiquitinated VDAC can be recognized by p62 to initiate mitophagy. DJ-1 senses oxidative stress and serves as a parallel pathway to maintain mitochondrial membrane potential and preserve mitochondria from fragmentation. Many of the regulators of mitophagy can be modulated by ROS. For example, a-synuclein is nitrated and, as a consequence, increases aggregation propensity. Parkin can be sulfonated and S-nitrosated. Drp-1 S-nitrosation is also involved in regulation of mitochondrial fission and associated induction of mitophagy. (From Lee et al., *Biochem. J.* 441:523–40, 2012. Copyright © 2012, The Biochemical Society. Reproduced with permission.)

18.4 SUMMARY

The benefits of mitophagy and pathogenic effects arising when mitophagy is dysfunctional are now being elucidated, and it is clear that degrading toxic mitochondria is likely a universal mechanism against oxidative and energy stress. Toxic mitochondria can arise by multiple mechanisms, a selected list of which follows:

- Mitochondria that have sustained an excessive level of damage to mtDNA and are beyond repair by fission/fusion
- Mitochondria with excessive damage of proteins of the electron transport chain, resulting in toxic levels of ROS
- Unwanted or excessive mitochondria that communally represent an oxidative threat to the cell

In the third scenario, "threat" can be envisioned to be caused by production of an excess of ROS by once healthy mitochondria that became toxic. This means that a healthy mitochondrion might be quickly switched to a toxic state. One explanation is that environmental conditions, especially energy homeostasis, which helps govern the rates of ROS production, have the signaling power to convert healthy mitochondria to toxic mitochondria. This raises the question of if or when a potentially toxic mitochondrion might be rescued and returned to a healthy state. Finally, mitophagy is a universal and essential mechanism protecting human mitochondria and their specialized, highly unsaturated mitochondrial membranes.

REFERENCES

Douarre, C., C. Sourbier, I. DallaRosa, et al. 2012. Mitochondrial topoisomerase I is critical for mitochondrial integrity and cellular energy metabolism. *PLoS One* 7:e41094. doi: 10.1371/journal.pone.0041094.

Fleming, A., T. Noda, T. Yoshimori, et al. 2011. Chemical modulators of autophagy as biological probes and potential therapeutics. *Nat. Chem. Biol.* 7:9–17.

Galluzzi, L., O. Kepp, C. Trojel-Hansen, et al. 2012. Mitochondrial control of cellular life, stress, and death. *Circ. Res.* 111:1198–207.

Green, D. R., L. Galluzzi, and G. Kroemer. 2011. Mitochondria and the autophagy-inflammation-cell death axis in organismal aging. *Science* 333:1109–12.

Kim, I., S. Rodriguez-Enriquez, and J. J. Lemasters. 2007. Selective degradation of mitochondria by mitophagy. *Arch. Biochem. Biophys.* 462:245–53.

Kurihara, Y., T. Kanki, Y. Aoki, et al. 2012. Mitophagy plays an essential role in reducing mitochondrial production of reactive oxygen species and mutation of mitochondrial DNA by maintaining mitochondrial quantity and quality in yeast. *J. Biol. Chem.* 287:3265–72.

Lee, J., S. Giordano, and J. Zhang. 2012. Autophagy, mitochondria and oxidative stress: crosstalk and redox signalling. *Biochem. J.* 441:523–40.

Levine, B., N. Mizushima, and H. W. Virgin. 2011. Autophagy in immunity and inflammation. *Nature* 469:323–35.

Pryde, K. R., and J. Hirst. 2011. Superoxide is produced by the reduced flavin in mitochondrial complex I: a single, unified mechanism that applies during both forward and reverse electron transfer. *J. Biol. Chem.* 286:18056–65.

Schweers, R. L., J. Zhang, M. S. Randall, et al. 2007. NIX is required for programmed mitochondrial clearance during reticulocyte maturation. *Proc. Natl. Acad. Sci. USA* 104:19500–5.

Suzuki, S. W., J. Onodera, and Y. Ohsumi. 2011. Starvation induced cell death in autophagy-defective yeast mutants is caused by mitochondria dysfunction. *PLoS One* 6:e17412. doi: 10.1371/journal.pone.0017412.

Yorimitsu, T., and D. J. Klionsky. 2005. Autophagy: molecular machinery for self-eating. *Cell Death Differ.* 12(Suppl 2):1542–52.

Youle, R. J., and D. P. Narendra. 2011. Mechanisms of mitophagy. *Nat. Rev. Mol. Cell Biol.* 12:9–14.

19 Longevity Genes Likely Protect Membranes

The race to identify human longevity genes has begun. A number of potential candidate genes have already been identified in several laboratories worldwide. This research in humans is at an early state and is dependent on developing methodology and criteria for identifying candidates for longevity genes. The next stage of research now underway involves testing candidate genes at the molecular or biochemical level as a means to verify their molecular role in enabling longevity. Thus far, some of the most promising candidates for longevity genes in humans are inspired by research on genes or gene variants accepted as being longevity genes in model animals (Kenyon, 2010), including *Caenorhabditis elegans*, fruit flies, and mice (see Section II). One of the most fertile grounds so far for finding potential longevity genes involves a universal regulatory pathway known as the insulin/insulin-like signaling pathway (Figure 19.1a). This highly sophisticated signaling cascade encoded by numerous genes and their regulatory elements ultimately governs expression of a vast array of genes, including energy metabolism, protection against oxidative stress, and energy stress.

19.1 ASHKENAZI JEWISH CENTENARIANS

Scientists at the Albert Einstein College of Medicine have carried out an extensive search for longevity genes among Ashkenazi Jewish centenarians and have found promising genetic variants in a gene encoding insulin-like growth factor 1 (IGF1) receptor (Suh et al., 2008). These researchers point out that complete or partial loss of function mutations in genes encoding components of the insulin/IGF1 pathway result in extension of life span in yeasts, worms, flies, and mice. They reason that the universality of this signaling mechanism across the animal world is a testament to its likely importance as a longevity gene in humans, specifically Ashkenazi Jewish centenarians. The availability of an extensive genome bank, along with a detailed record of lifestyle of many aged individuals and their offspring, is an important advantage for this research. Interestingly, it has been publicized that some of the aged individuals in this study engaged throughout life in activities that might seem the antithesis of healthy living. One scenario derived from these data on behavior is that genes might sometimes trump environment for a long, healthy life.

From a single receptor in *C. elegans* the insulin/IGF1 pathway in humans is very complex, with multiple receptors and a far more complicated regulatory network. Insulin regulates many metabolic pathways, in contrast to IGF1 and growth hormone, which control growth and differentiation. Previous extensive mutational analysis of these pathways has been conducted, in some cases giving opposing conclusions. Recent data suggest that partial loss of function mutants in human IGF1, caused by single nucleotide changes, enhance life span (Suh et al., 2008). Tazearslan and colleagues (2011) have located these amino acid substitutions placed in the context of the crystal structure of appropriate segments of insulin-like growth factor 1 receptor (IGF1R).

To further evaluate the role of these putative human longevity genes, the Albert Einstein research group has constructed cell lines expressing genes encoding two different IGF1R variants (Ala-37-Thr and Arg-407-His) (Tazearslan et al., 2011). In this study mouse embryonic fibroblasts (MEFs) were engineered to express the different IGF1R variants. The data show that MEFs expressing the human longevity-associated IGF1R mutations attenuated IGF1 signaling, as demonstrated by significant reduction in phosphorylation of IGF1R after IGF1 treatment in comparison with MEFs expressing the wild-type IGF1R. The impaired IGF1 signaling caused by the IGF1R mutations resulted in the reduced induction of the major IGF1-activated genes in cells. The IGF1R mutations also caused a delay in cell cycle progression after IGF1 treatment, indicating a dysfunctional physiological response to a cell proliferation signal. These results demonstrate that the human longevity-associated IGF1R variants are reduced-function mutations, implying that dampening of IGF1 signaling may be a longevity mechanism in humans.

Integration of the IGF1 signaling pathway and the reactive oxygen species (ROS)-mediated signaling pathways for protecting membranes involves regulatory cross talk between these pathways. The molecular mechanisms of cross talk may play an important role in the mechanism of regulation of aging and longevity. Mouse models have been developed for both ROS-mediated and IGF1 signaling pathways (see review by Papaconstantinou, 2009). A common physiological characteristic shared by genetic models of delayed aging and extended life span includes increased resistance to oxidative stress and protection of highly unsaturated mitochondrial membranes.

19.2   FOXO TRANSCRIPTION FACTORS GOVERN LONGEVITY IN MODEL ANIMALS AND PERHAPS HUMANS

Discovery of a potential longevity gene passed from generation to generation called FOXO3A in human centenarians has created major repercussions throughout the field of aging (Willcox et al., 2008; Flachsbart et al., 2009), including a revised holy grail of aging. These data raise a new standard for a healthy human life span to a century. These data also support the view, contrary to official Food and Drug Administration (FDA) doctrine, that aging might be a disease to be cured.

The history of research on FOXO3A transcription factors in extending longevity in humans is summarized as follows:

- 1993: Mutant FOXO doubles the life span of wild-type *C. elegans.*
- 2002: FOXO found to play critical roles in cell cycle and death control.
- 2004: FOXO role in cellular metabolism, differentiation established.
- 2005: Molecular biology of FOXO transcription factors elucidated and linkage to longevity proposed.
- 2005: FOXO role in longevity of *C. elegans* generalized to other animals.
- 2005: Unique patterns of expression of multiple forms of FOXO determined in mouse brain.

- 2006: Linkage of FOXO to oxidative stress and life span described.
- 2007: FOXO found to mediate stem cell resistance to oxidative stress.
- 2008: Unified concept of FOXO transcription factors in maintaining cellular homeostasis during aging proposed.
- 2008: FOXO3A genotype found to be strongly associated with human longevity.
- 2009: FOXO shown to regulate neural stem cell homeostasis.
- 2009: FOXO3A link to longevity in centenarians confirmed.
- 2011: Detailed integration and analysis of data on FOXO3A raises some controversial issues and suggests some revisions on the scope of this system are in order.

Theoretically, the specialized forms and patterns of expression of FOXO transcription factors in cells seem ideally suited to enable long-term protection of membranes. This hypothesis makes sense from a bioenergetic perspective because lifelong protection of membranes is expected to save a great deal of energy by at least two mechanisms. First, FOXO transcription factors are likely timed to coincide with periods of need, reducing the long-term metabolic cost associated with constitutive synthesis of this protective system. Second, there might be both direct and indirect energy savings associated with protecting membranes. FOXO expression, and thus levels of protection, is likely to be continually up-regulated to counterbalance any increases in age-related stresses caused by aging itself. Age-related stresses likely include oxidative, energy, and toxic peptide stresses. Interestingly, the DHA principle predicts that neurons require extraordinary lifelong protection against oxidative stresses and energy stresses, a topic we have recently discussed (Valentine and Valentine, 2013).

The molecular biology of FOXO transcription factors is understood in considerable detail (Figure 19.1b) (Accili and Arden, 2004; Burgering and Kops, 2002; Greer and Brunet, 2005; Hoekman et al., 2006; Lehtinen et al., 2006; Tothova et al., 2007, Miyamoto et al., 2007; Salih and Brunet, 2008; Paik et al., 2009, Renault et al., 2009). A diverse set of genes is modulated by FOXO factors and includes genes essential for bioenergetics, development, cell cycle, death, and protection. It is especially interesting that FOXO factors up-regulate genes essential for long-term health of all human cells. Indeed, the protective effect of the FOXO system seen through the lens of molecular biology seems powerful enough to increase human longevity.

FOXO transcription factors are modulated by elaborate regulatory cascades that sense and respond to environmental signals, including signals coming from both outside and inside the cell (Figure 19.1b). FOXO factors change their structures in response to changes in the world around them (Greer and Brunet, 2005). In essence, the most sophisticated kinds of regulatory circuitry or biochemistry in nature are seen in this family of regulatory proteins. FOXO factors are regulated by environmental signals as diverse as hormones, glucose levels, or oxidative stress. It is little wonder that genetic variations in these regulatory proteins result in a mind-boggling number of phenotypes, including genetic variants that increase longevity in model animals and perhaps humans. FOXO proteins reside and are active in the nucleus where they bind to DNA as monomers via the Forkhead domain or box, which is a 110-amino acid region located roughly in the central position of the molecule. When present in the nucleus and attached to DNA, FOXO regulators behave as potent

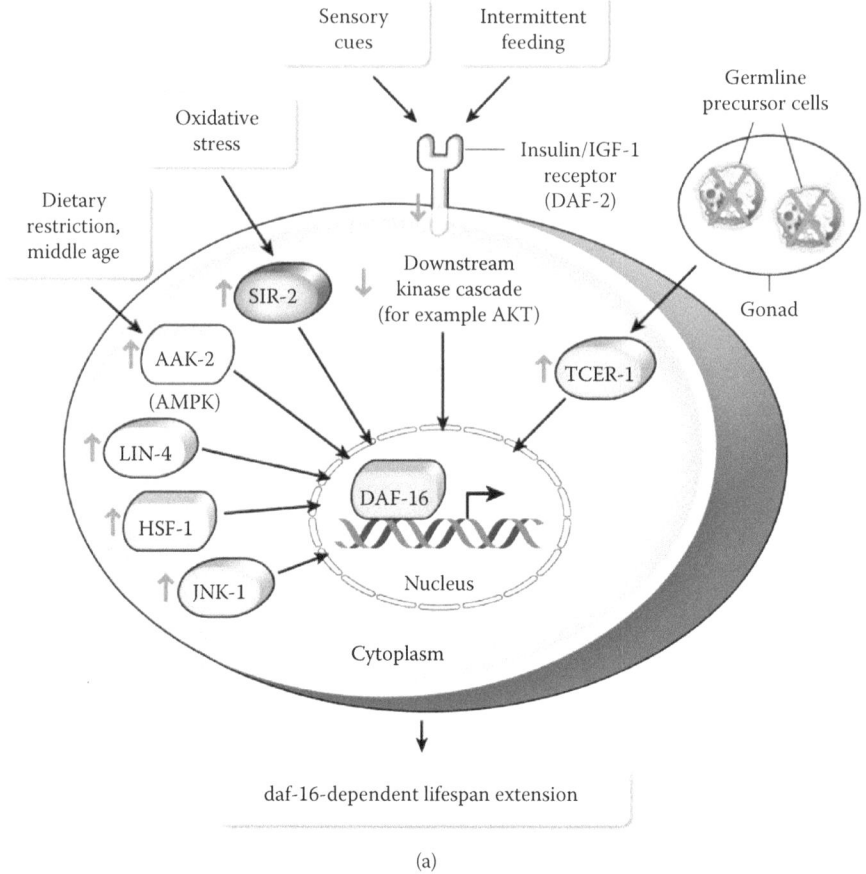

(a)

FIGURE 19.1 **See color insert.** DAF-16 (FOXO-like transcription factor of *C. elegans*) regulates stress responses in *C. elegans*. (a) Overexpressing DAF-16, SIR-2 (sirtuin), HSF-1 (heat shock transcription elongation factor), LIN-4 (developmental-tuning micro RNA), AAK-2 (a subunit of AMP kinase), JNK-1 (JUN kinase), or the transcription elongation factor TCER-1 extends life span. Inhibiting the DAF-2 insulin/IGF1 receptor or components of its downstream kinase cascade also extends life span. In each case the extension of life span is DAF-16 dependent. (From Kenyon, *Nature* 464:504–12, 2010. Copyright © 2010. Reprinted by permission from Macmillan Publishers Ltd.) *(continued)*

transcriptional activators up-regulating synthesis of specific messenger RNAs, likely including transcripts that are translated into stress-protective proteins. The activation domain of FOXO factors is located in the C-terminal portion of the protein. FOXO proteins can switch from being transcriptional activators to repressors of gene activity, depending on the specific DNA promotor to which they bind, along with existing environmental conditions. Thus, the presence of FOXO proteins in all human cell types, along with their exquisite ability to sense and respond to changes in their surroundings, makes them ideal candidates for pushing back against aging. The different homologs of FOXO factors that have evolved in humans, along with their unique

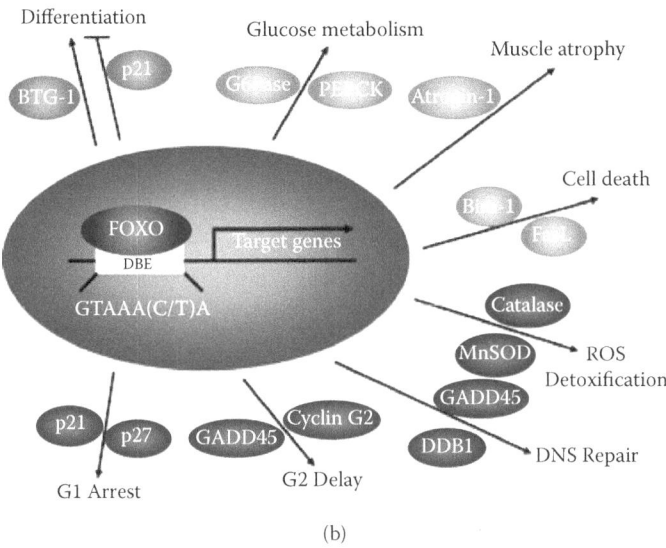

(b)

FIGURE 19.1 *(continued)* **See color insert.** DAF-16 (FOXO-like transcription factor of *C. elegans*) regulates stress responses in *C. elegans*. (b) Multiple FOXO transcription factors govern stress tolerance in mammals. (From Greer and Brunet, *Oncogene* 24:7410–25, 2005. Copyright © 2005. Reprinted by permission from Macmillan Publishers Ltd.)

expression patterns seen in different tissues, especially the brain, point to organ-specific, cellular-specific, or even life cycle-specific protection by these transcription factors. For example, FOXO6 is predominantly expressed in the developing brain, whereas FOXO3 is expressed in many cells and highly and constitutively expressed in brain tissue, including the adult brain. Similar specific patterns of expression have been seen with various FOXO transcription factors in other tissues. Data from the pattern of expression of FOXO factors in the brain are consistent with a specialized need for protection of neurons during development, followed by unusually high levels of protection over a lifetime. This pattern of expression is also consistent with the DHA principle applied to aging that states that DHA/polyunsaturated membranes require specialized or tailor-made protection systems that must remain strong as we age. Thus, molecular studies of FOXO factors are expected to have major impacts on understanding both basic and applied aspects of aging.

We suggest that the elegant data on the molecular biology of FOXO factors are consistent with the dual-energy pacemaker model of cellular health, in which energy stress worsens with aging. Theoretically, FOXO factors are among the most important long-term energy savers and energy boosters yet discovered in animals. The energy savings come from steadily up-regulating protective systems as they weaken with aging, thus protecting cellular membranes and proteins from oxidative damage. In essence, prolonging the functional life span of membranes and other key cellular macromolecules, structures, and organelles, including mitochondria, has both direct and indirect impacts on oxidative stress and energy homeostasis. We have emphasized in this book that oxidatively damaged membranes and membrane proteins can directly cause leakage or uncoupling of essential ion gradients, especially bioenergetically important gradients of protons in mitochondria and sodium in axons. We suggest that energy uncoupling at the membrane level is far more important in aging than currently appreciated (see Chapter 16). The indirect effect comes in the form of less energy being spent to constantly repair mitochondria, damaged membranes, proteins, and DNA. The direct effect involves the energy cost associated with purposeful uncoupling of proton gradients by UCPs (Chapter 16). Differential expression of FOXO factors likely evolved as a lifelong insurance policy against stresses, a mechanism especially important in preventing oxidative damage of membranes.

Mitochondria are considered a major pacemaker of neurodegeneration. The FOXO family might allow aging neuron mitochondria to work more slowly, longer, and more efficiently, prolonging the effective period of energy production before mitochondria require repair. Reducing the workload of mitochondria in the face of oxidative damage might allow these organelles to produce energy for a longer period of time and more efficiently and, as a result, slow down the mtDNA pacemaker of aging.

19.3   HYPOTHESIS THAT NEURONS ACT AS MASTER REGULATORS OF AGING AND AS A SOURCE OF LONGEVITY GENES, INCLUDING APOLIPOPROTEIN D (APOD)

Data from studies of animal models of aging suggest that neurons might play a role as a central integrator of physiological changes linked to aging (see review by Bishop et al., 2010). It is clear that aging is the major risk factor for neurodegeneration, which means that cascades governing human brain span and life span are linked. The above authors take the brain-body relationship to a higher level, proposing that genes that enable a healthy brain span might act as generalized longevity genes, with neurons being a rich source of these genes (Bishop et al., 2010). These researchers propose a network of regulatory circuitry allowing neurons of the brain to communicate and cross talk with other organs during the aging process. This concept not only puts the brain in the position of sensing that aging is occurring within itself and the rest of the body, but also suggests that the brain has the power to govern the pace of aging. According to this model, higher-order brain systems become less efficient with aging, perhaps due to gene expression changes that alter fundamental neural function, including myelination, axonal integrity, and synaptic function. Membranes and their functional state play cardinal roles in these processes. Bishop and coworkers (2010) raise the point that age-related dysfunction in the brain might lead to disruption of a putative brain-body feedback loop, perhaps involving crucial hormonal systems.

A unified mechanism of brain aging and general aging along with a high level of cross talk between the brain and body leads to the intriguing idea that neurons might be a source of longevity genes. Genes involved in stress response pathways are universally important during aging, especially in the brain. One gene that has received considerable attention encodes apolipoprotein D. This gene is up-regulated in the aging cortex of mice, rhesus macaques, and humans (Loerch et al., 2008). Importantly, lipoprotein D expression extends life span in *Drosophila*, and the protein functions as a lipid antioxidant conferring resistance to oxidative stress (Walker et al., 2006). Whereas up-regulation of ApoD in *Drosophila* results in increased life span, the opposite occurs in loss of function mutants for the GLaz gene, a homolog of ApoD. The absence of GLaz reduces the organism's resistance to oxidative stress, including paraquat, and these mutants have a higher concentration of lipid peroxidation products. These data suggest protection of membranes against oxidative damage is a possible mechanism of action. Human ApoD (hApoD), which is up-regulated in the human brain during aging, has been expressed in *Drosophila* and been found to increase stress resistance (Muffat et al., 2008). Flies overexpressing hApoD are long-lived and protected against stress conditions, including oxidative stress mediated by paraquat. Among its many functions, ApoD may be involved in the clearance or repair of oxidatively damaged membranes. ApoD, a small protein, is up-regulated 500-fold at the site of the sciatic nerve crush injury in the rat. It is important to note that regulation of ApoD levels is governed by toxic compounds in the environment, such as paraquat, as well as intrinsic signals, including aging and other shifts in metabolism.

GLaz in *Drosophila* and ApoD in humans belong to a family of proteins called lipocalins that bind hydrophobic molecules, including lipids and their by-products. Lipocalins play numerous physiological roles, including lipid transport, anti-inflammatory and antimicrobial function, and importantly, scavenging of toxic molecules such as lipid peroxidation products (Lechner et al., 2001). Some lipocalins have enzymatic properties, including antioxidant activities (Allhorn et al., 2005; Lechner et al., 2001). Thus, the lipocalin alpha-1-microglobin is able to protect tissues from preoxidant molecules, inhibit oxidation, and remove oxidation products.

Recently, a large screening for human longevity genes was carried out (Walter et al., 2011) among 25,000 individuals of fifty-five years and older followed for an average of eleven years. The main result is that several loci with suggestive significance for survival free of major diseases were found in the neighborhood of genes related to neural regulation. These researchers identified genes associated with cellular and neural development and function and cell communication that may contribute to variation in human aging. Brain function is essential for modulating cellular homeostatic processes throughout the body, including energy homeostasis and oxidative stress homeostasis. Thus, healthy neurons with healthy membranes are essential for disease-free aging and longevity.

Walter and colleagues (2011) show that ten of the twenty-two suggestive associations identified in their analyses are in or near genes that are highly expressed in the brain. These genes were previously linked to the regulation of neuronal excitability and plasticity and the maintenance of neural circuitry and synaptic plasticity. Several genes captured as being candidates for longevity genes are associated with neurological diseases, including Alzheimer's disease and amyotrophic lateral sclerosis. In addition, six of the single nucleotide polymorphisms (SNPs) were in close proximity to genes linked with other phenotypes of aging, including autophagy, cancer, and mitochondrial depletion syndrome.

Arachidonic acid (ARA) (20:4) is found in fish oil and is classified as an omega-6 fatty acid and as a highly unsaturated fatty acid (HUFA) with four double bonds. ARA is present in significant amounts in axon phospholipids of neurons and is distributed widely in plasma membranes of most human cells. The phospholipid form of ARA is cleaved by phospholipase to yield the free acid. A low amount of the free acid is present in brain tissue. Arachidonic acid is best known as a precursor for a series of potent eicosanoid hormones. Eicosanoids are needed only in trace amounts and are highly unstable. In the brain, eicosanoids are produced and function locally. As hormones, eicosanoids play a variety of roles, including acting as triggers of inflammation. The linkage between eicosanoids and inflammation was established during studies of the mode of action of common aspirin (acetylsalicylic acid) acting as a painkiller. The pain cascade is blocked after the acetyl moiety of aspirin is transferred to and blocks the active site of the eicosanoid-producing enzyme cyclooxygenase. Thus, aspirin stops pain and inflammation by blocking the first step in the enzymatic synthesis of eicosanoids, which shows that eicosanoids act in a cascade to trigger pain and inflammation.

For long-term brain health, too much eicosanoid production may be harmful, essentially creating a state of chronic inflammation, which is a form of oxidative stress. This is where the small, highly regulated apolipoprotein D (ApoD) protein molecule might come into the picture. The structure of ApoD shows a deep binding pocket that is highly specific for certain lipids—the steroid hormone progesterone and arachidonic acid (Eichinger et al., 2007). A single ARA molecule fits tightly in the pocket such that ApoD has a powerful ability to sequester and remove ARA from the brain. Free ARA is present in low amounts in brain tissue, though its levels may rise or fall with aging, disease, or after an oxidative challenge. It is interesting to speculate that a regulatory system allowing dramatic fluctuation of expression of ApoD evolved in the brain as a mechanism to sequester ARA and modulate its levels below a critical threshold that can cause neurodegeneration. The main point is that too much ARA likely creates a state of persistent inflammation-oxidative stress that is not compatible with a long brain span. We hypothesize that ApoD, like aspirin, blocks the inflammatory cascade caused by ARA, thus decreasing oxidative stress and membrane peroxidation in the brain, a pro-longevity mechanism.

The following data (Elliott et al., 2010; Eichinger et al., 2007; Ganfornina et al., 2008; Muffat et al., 2008; Sanchez et al., 2006) are consistent with this model:

- ApoD is ubiquitous in distribution and function and highly regulated among animals, ranging from flies to humans (e.g., 500-fold increase in ApoD levels during smash-induced neuron damage in rats).
- Loss of function mutants of ApoD in *Drosophila* decrease longevity, whereas overexpression increases longevity.
- Human ApoD overexpression in recombinant *Drosophila* increases longevity.
- Membrane peroxidation levels are predicted by levels of ApoD, with membrane damage being low when ApoD levels are elevated, and vice versa.
- Biochemical mode of action of ApoD is consistent with tight sequestering of ARA to block conversion of ARA → eicosanoids.

Evidence linking ApoD with Alzheimer's disease (AD) (reviewed by Elliott et al., 2010; see their extensive reference list) is discussed next. In AD, increased levels of ApoD are found in cerebrospinal fluid (CSF) and in diseased regions of the brain, including the hippocampus, frontal cortex, and temporal cortex. ApoD immunostaining of cortical neurons suggests a correlation with the presence of neurofibrillary tangle density (a signature of AD pathology). ApoD levels are increased in the parkinsonian region of the brain during Parkinson's disease and in CSF of patients with meningoencephalitis, stroke, dementia, and motor neuron disease. In an animal model of the human cholesterol storage disorder Niemann-Pick disease, ApoD levels were found to increase about 30-fold. Recent data support a generalized role of ApoD in preventing membrane peroxidation, especially important for maintaining brain membrane homeostasis.

Ultimately, protecting neurons, specifically neuron membranes, is likely an important mechanism enabling long brain span. An alternative model has recently been proposed in which the protective role of ApoD occurs at the level of astrocytes (Bajo-Grañeras et al., 2011). Clearly, protecting astroglial cells against oxidative stress is important, as is protecting neurons; protecting astrocytes against oxidative stress may also protect neurons. Another explanation is that ApoD plays multiple roles in the brain, protecting both astrocytes and neurons.

19.4 GENETIC VARIANTS OF UNCOUPLING PROTEINS (UCPS) AS LONGEVITY GENES

A set of five human genes, encoding uncoupling proteins (UCPs 1–5) and found in the inner mitochondrial membrane (Chapter 16), are receiving increasing attention as longevity-enhancing genes. Uncoupling proteins that modulate bioenergetics and regulation of ROS production have been manipulated in *Drosophila* (Fridell et al., 2009) and mice (Speakman et al., 2004; Conti et al., 2006; Andrews and Horvath, 2009) to increase longevity. Interestingly, transgenic mice with a reduced core body temperature have an increased life span (Conti et al., 2006).

Recently genetic variations of human UCP genes have been found to be associated with longevity (Rose et al., 2011). Genetic variability of human UCP found in a total of 598 subjects ranging between 64 and 100 years of age showed that genetic differences of UCP2, UCP3, and UCP4 affect the individual's chances of surviving up to a very old age. This supports the importance of energy storage, energy use, and modulation of ROS production in the aging process. Given the expression of UCP2 in brain, heart, and adipose tissue, UCP3 in muscles and brown adipose tissue, and UCP4 in neuronal tissue, these data suggest that the energy uncoupling process, which can down-regulate ROS production by mitochondria, plays an important role in modulating aging (see Chapter 16).

19.5 SUMMARY

Among the human longevity gene candidates discussed above, several involve mechanisms that might enable long-term prevention of oxidative damage in membranes. FOXO and IGFR1 are both part of the same insulin/insulin-like signaling pathway, which means that any roles they play in increasing life span are linked to their mode of action as genetic regulators. That is, this signaling system acts by raising or lowering levels of protective enzymes or proteins. In the case of membrane protection, the ability to increase or decrease a diverse battery of protective enzymes would seem to have a powerful selective advantage, especially in long-lived humans. The fact that genetic variations in members of the insulin/insulin-like signaling system confer longevity in animal models has spurred a great interest in similar variants in humans. Research on this topic in humans began in earnest after the report that a variant of FOXO is present in the genome of centenarians. A strong literature on the molecular biology of FOXO transcription factors provides a solid foundation for the FOXO system as applied to humans, a developing story.

From the perspective of mechanisms protecting membrane fatty acids against peroxidation, the ApoD system is intriguing, with implications for explaining linkages among inflammation, membrane peroxidation, and aging. Recent data show that manipulating uncoupling proteins (UCPs) in *Drosophila*, mice, and humans can increase longevity. UCPs seem to bridge the gap between mitochondrial bioenergetics and production of and prevention of oxidative damage of membranes by reactive oxygen species. Finally, the current trajectory of research on longevity genes is trending toward recognition that membrane lipid molecular biology and aging are linked. The common denominator of this linkage is the need to prevent membrane peroxidation, perhaps enabling longevity. A few setbacks should not discourage the basic and applied research community from teaming up toward understanding and managing aging, though this field is still in its infancy.

REFERENCES

Accili, D., and K. C. Arden. 2004. FoxOs at the crossroads of cellular metabolism, differentiation, and transformation. *Cell* 117:421–26.

Allhorn, M., A. Klapyta, B. Akerström, et al. 2005. Redox properties of the lipocalin alpha1-microglobulin: reduction of cytochrome c, hemoglobin, and free iron. *Free Radic. Biol. Med.* 38:557–67.

Andrews, Z. B., and T. L. Horvath. 2009. Uncoupling protein-2 regulates lifespan in mice. *Am. J. Physiol. Endocrinol. Metab.* 296:E621–27.

Bajo-Grañeras, R., M. D. Ganfornina, E. Martín-Tejedor, et al. 2011. Apolipoprotein D mediates autocrine protection of astrocytes and controls their reactivity level, contributing to the functional maintenance of paraquat-challenged dopaminergic systems. *Glia* 59:1551–66.

Bishop, N. A., T. Lu, and B. A. Yankner. 2010. Neural mechanisms of ageing and cognitive decline. *Nature* 464:529–35.

Burgering, B. M., and G. J. Kops. 2002. Cell cycle and death control: long live Forkheads. *Trends Biochem. Sci.* 27:352–60.

Conti, B., M. Sanchez-Alavez, R. Winsky-Sommerer, et al. 2006. Transgenic mice with a reduced core body temperature have an increased life span. *Science* 314:825–28.

Eichinger, A., A. Nasreen, H. J. Kim, et al. 2007. Structural insight into the dual ligand specificity and mode of high density lipoprotein association of apolipoprotein D. *J. Biol. Chem.* 282:31068–75.

Elliott, D. A., C. S. Weickert, and B. Garner. 2010. Apolipoproteins in the brain: implications for neurological and psychiatric disorders. *Clin. Lipidol.* 51:555–73.

Flachsbart, F., A. Caliebe, R. Kleindorp, et al. 2009. Association of FOXO3A variation with human longevity confirmed in German centenarians. *Proc. Natl. Acad. Sci. USA* 106:2700–5.

Fridell, Y. W., M. Hoh, O. Kréneisz, et al. 2009. Increased uncoupling protein (UCP) activity in *Drosophila* insulin-producing neurons attenuates insulin signaling and extends lifespan. *Aging (Albany NY)* 1:699–713.

Ganfornina, M. D., S. DoCarmo, J. M. Lora, et al. 2008. Apolipoprotein D is involved in the mechanisms regulating protection from oxidative stress. *Aging Cell* 7:506–15.

Greer, E. L., and A. Brunet. 2005. FOXO transcription factors at the interface between longevity and tumor suppression. *Oncogene* 24:7410–25.

Hoekman, M. F. M., F. M. J. Jacobs, M. P. Smidt, et al. 2006. Spatial and temporal expression of FoxO transcription factors in the developing and adult murine brain. *Gene Express. Patt.* 6:134–40.

Kenyon, C. J. 2010. The genetics of ageing. *Nature* 464:504–12.

Lechner, M., P. Wojnar, and B. Redl. 2001. Human tear lipocalin acts as an oxidative-stress-induced scavenger of potentially harmful lipid peroxidation products in a cell culture system. *Biochem. J.* 356:129–35.

Lehtinen, M. K., Z. Yuan, P. R. Boag, et al. 2006. A conserved MST-FOXO signaling pathway mediates oxidative-stress responses and extends life span. *Cell* 125:987–1001.

Loerch, P. M., T. Lu, K. A. Dakin, et al. 2008. Evolution of the aging brain transcriptome and synaptic regulation. *PLoS One* 3:e3329.

Miyamoto, K., K. Y. Araki, K. Naka, et al. 2007. Foxo3a is essential for maintenance of the hematopoietic stem cell pool. *Cell Stem Cell* 1:101–12.

Muffat, J., D. W. Walker, and S. Benzer. 2008. Human ApoD, an apolipoprotein up-regulated in neurodegenerative diseases, extends lifespan and increases stress resistance in *Drosophila*. *Proc. Natl. Acad. Sci. USA* 105:7088–93.

Paik, J. H., Z. Ding, R. Narurkar, et al. 2009. FoxOs cooperatively regulate diverse pathways governing neural stem cell homeostasis. *Cell Stem Cell* 5:540–53.

Papaconstantinou, J. 2009. Insulin/IGF-1 and ROS signaling pathway cross-talk in aging and longevity determination. *Mol Cell. Endocrinol.* 299:89–100.

Renault, V. M., V. A. Rafalski, A. A. Morgan, et al. 2009. FoxO3 regulates neural stem cell homeostasis. *Cell Stem Cell* 5:527–39.

Rose, G., P. Crocco, F. DeRango, et al. 2011. Further support to the uncoupling-to-survive theory: the genetic variation of human UCP genes is associated with longevity. *PLoS One* 6:e29650.

Salih, D. A., and A. Brunet. 2008. FoxO transcription factors in the maintenance of cellular homeostasis during aging. *Curr. Opin. Cell Biol.* 20:126–36.

Sanchez, D., B. López-Arias, L. Torroja, et al. 2006. Loss of glial lazarillo, a homolog of apolipoprotein D, reduces lifespan and stress resistance in *Drosophila*. *Curr. Biol.* 16:680–86.

Speakman, J. R., D. A. Talbot, C. Selman, et al. 2004. Uncoupled and surviving: individual mice with high metabolism have greater mitochondrial uncoupling and live longer. *Aging Cell* 3:87–95.

Suh, Y., G. Atzmon, M. O. Cho, et al. 2008. Functionally significant insulin-like growth factor I receptor mutations in centenarians. *Proc. Natl. Acad. Sci. USA* 105:3438–42.

Tazearslan, C., J. Huang, N. Barzilai, et al. 2011. Impaired IGF1R signaling in cells expressing longevity-associated human IGF1R alleles. *Aging Cell* 10:551–54.

Tothova, Z., R. Kollipara, B. J. Huntly, et al. 2007. FoxOs are critical mediators of hematopoietic stem cell resistance to physiologic oxidative stress. *Cell* 128:325–39.

Valentine, R. C., and D. L. Valentine. 2013. *Neurons and the DHA principle*. Boca Raton, FL: Taylor and Francis Group.

Walker, D. W., J. Muffat, C. Rundel, et al. 2006. Overexpression of a *Drosophila* homolog of apolipoprotein D leads to increased stress resistance and extended lifespan. *Curr. Biol.* 16:674–79.

Walter, S., G. Atzmon, E. W. Demerath, et al. 2011. A genome-wide association study of aging. *Neurobiol. Aging* 32:2109.e15–28.

Willcox, B. J., T. A. Donlon, Q. He, et al. 2008. FOXO3A genotype is strongly associated with human longevity. *Proc. Natl. Acad. Sci. USA* 105:13987–92.

20 Aging as a Cardiolipin Disease That Can Be Treated

As background for our final statement, we return full circle to the Global Burden of Disease Study that states that a majority of patients in the future will require treatment for age-dependent diseases, especially ailments of an aging brain and nervous system (see Section I). Currently, many of these diseases are untreatable, and in the case of aging, the Food and Drug Administration (FDA) of the United States has yet to approve a single drug treatment. Indeed, according to the FDA, aging is not a disease to be cured. Thus, we are faced with a quandary since results of the Global Burden of Disease Study imply the opposite—many ailments of aging are diseases that need to be cured and will dominate healthcare in the twenty-first century. In this chapter we go a step further and evaluate evidence showing that age-dependent diseases and aging itself share much in common and might be treated by common drugs. That is, the molecular pathologies associated with aging and age-dependent diseases seem to converge on the same point. According to this unified concept, advances in understanding or treatments for age-dependent disease such as Parkinson's disease will likely apply to other ailments of aging and vice versa.

20.1 WORKING DEFINITION OF CARDIOLIPIN DISEASES

Mitochondria-targeted antioxidants are a specialized class of antioxidants designed to protect mitochondria against oxidative damage. So far this class of drugs has had only limited success in treatment of human diseases, but a new generation of mitochondria-targeted antioxidants is sparking excitement in the field of neurodegeneration and age-dependent diseases. For a review of mitochondria-targeted antioxidants used previously in human drug trials, see Smith and Murphy (2011), and for a review of the linkage of mitochondrial bioenergetics and Alzheimer's disease, see Calkins and colleagues (2012). Recently Ji and colleagues (2012) reported a critical test of the cardiolipin hypothesis of age-dependent diseases with implications for aging. These important data have two major consequences. First, these data provide the most compelling evidence yet that cardiolipin oxidation triggers neurodegeneration in rodent models. A causal relationship between cardiolipin and neurodegeneration was established using a mitochondria-targeted antioxidant that blocks oxidation of cardiolipin while preventing neurodegeneration. The second contribution is that these data illuminate the potential of mitochondria-targeted antioxidants in treatment of many human diseases. These data also suggest that many catastrophic and untreatable human diseases might be defined in terms of a cluster of diseases with cardiolipin as the trigger—cardiolipin diseases. A second class of antioxidants targeting membranes in general, but especially mitochondrial phospholipids (including cardiolipin), has been shown in rodent models to protect parkinsonian neurons against oxidative death.

20.2 DEFINITIVE PROOF THAT DOUBLE BONDS OF POLYUNSATURATED MEMBRANE PHOSPHOLIPIDS CAN CAUSE OXIDATIVE DEATH OF CELLS

In Silicon Valley, California, a start-up company called Retrotope, Inc. is pioneering a new class of membrane-targeted drugs to treat neurodegenerative diseases (Hill et al., 2011, 2012; Shchepinov et al., 2011). A specific bioactive fatty acid derivative called deuterated linoleic acid (D-18:2) is found in small amounts in nature, but Retrotope, Inc. uses a chemically pure, deuterated derivative of D-18:2 in which D replaces a specific, facile hydrogen along the chain. By substituting deuterium in place of hydrogen on a specific location on the unsaturated fatty acid chain, Retrotope scientists are able to prevent oxidative damage from triggering a deadly chain reaction. These researchers initially used a yeast model to show for the first time that double bonds of 18:2 trigger oxidative death of yeast cells.

Yeast cells whose membranes are enriched with polyunsaturated fatty acids (PUFAs) such as 18:2 replacing monounsaturated fatty acids (MUFAs) such as 18:1 (Walenga and Lands, 1975) are known to be sensitized to killing by oxidative stress (Do et al., 1996). Lethality might be caused by increased rates of peroxidation of PUFAs triggering catastrophic oxidative damage to membranes and other cellular constituents. The previous correlative evidence supporting this model in PUFA-enriched yeast cells (which normally don't produce polyunsaturated fatty acids) is as follows:

- O_2 is required for killing of 18:2-enriched cells.
- Copper, a known catalyst for membrane peroxidation, greatly enhances cellular death.
- 18:3-enriched cells are more sensitive to copper-mediated killing than 18:2, with the trend 18:1 < 18:2 < 18:3 in increasing sensitivity.
- Reactive oxygen species (ROS) levels are elevated in PUFA-enriched cells.

These data are consistent with a mechanism in which peroxidation of PUFAs causes cellular death. Do and coworkers (1996) also reported that ubiquinone-deficient mutants of *Saccharomyces cerevisiae* are especially sensitive to PUFAs. This hypersensitivity of yeast to a genetic block in ubiquinone synthesis (Do et al., 1996) was initially explained by a role of reduced ubiquinone acting directly as an antioxidant. However, a possible alternative mechanism has now emerged. Solution of the structure of complex 1 of the respiratory chain (see Chapters 11 and 16) provides new insight into how ubiquinone-deficient mutants of yeast might become hypersensitive to killing by dietary PUFAs. According to the flavin site mechanism of complex 1 developed by Pryde and Hirst (2011), any event that raises or lowers the mitochondrial NAD^+ pool may effect ROS production, which is closely linked to oxidative damage of mitochondrial membranes. Thus, a deficiency of ubiquinone is expected to decrease electron flow along the electron transport system of yeast mitochondria and increase NADH levels, resulting in hyperreduction of the high-energy electron site of complex 1. The net effect is formation of excessive amounts of ROS by complex 1, ultimately lethally damaging membranes. Death of a PUFA-enriched yeast cell might be the result of both an oxidative chain reaction propagated by PUFAs replacing MUFAs in the inner mitochondrial membranes and the loss of coenzyme Q acting as an antioxidant.

Recently scientists associated with Retrotope applied knowledge of the hallmark chemistry involving initiation and propagation of membrane peroxidation reactions to show definitively that PUFAs trigger lethal oxidative stress in yeast and how deuterated PUFAs incorporated into membrane phospholipids protect cells against oxidative death (Hill et al., 2011, 2012). These authors show that yeast cells enriched with polyunsaturated fatty acids deuterated at bis-allylic sites (Figure 20.1a) are protected against oxidative stress because of the isotope effect (Figure 20.1b) (Hill et al., 2011, 2012). The chemical rationale for the isotope effect on PUFA peroxidation is that the rate of reaction involving C-H bond cleavage is typically five to ten times faster than the corresponding C-D bond cleavage, due to the two-fold difference in the masses of H versus D. Deuterium effects on enzymatic lipoxidation reactions serve as a model to show that hydrogen abstraction is usually the rate-limiting step. For example, the reaction of soybean lipoxygenase-1 with its substrate, linoleic acid (18:2), displays very large kinetic isotope effects, predicting similar effects in membranes (Glickman and Klinman, 1995). The data of Hill and colleagues (2011, 2012) are a milestone in biochemistry of membranes since they demonstrate a causal relationship between oxidation of double bonds of PUFAs and cellular death. A model developed by Retrotope to explain how PUFAs trigger a lethal chain reaction is shown in Figure 20.2. The above models also explain how deuteration prevents both chemical peroxidation and enzymatic lipoxidation of cardiolipin.

20.3 DEUTERATED 18:2 (D-18:2) PROTECTS NEURONS AGAINST OXIDATIVE DEATH IN A MOUSE MODEL OF PARKINSON'S DISEASE

Isotope protection of yeast against polyunsaturation-mediated oxidative killing suggests a new approach to preventing ROS-induced membrane damage during aging and age-dependent diseases (Shchepinov, 2007; Shchepinov et al., 2011). For example, Parkinson's disease is thought to be caused in part by oxidative damage of mitochondrial membranes generating catastrophic oxidative damage in parkinsonian cells of the brain. Shchepinov and coworkers (2011) have tested this hypothesis with the result that deuteration provides isotopic protection against oxidative damage and provides partial protection against nigrostriatal injury in a mouse model of Parkinson's disease. The neurotoxin MPTP was used as a positive control in these experiments, where, as expected, this potent parkinsonian chemical, like paraquat,

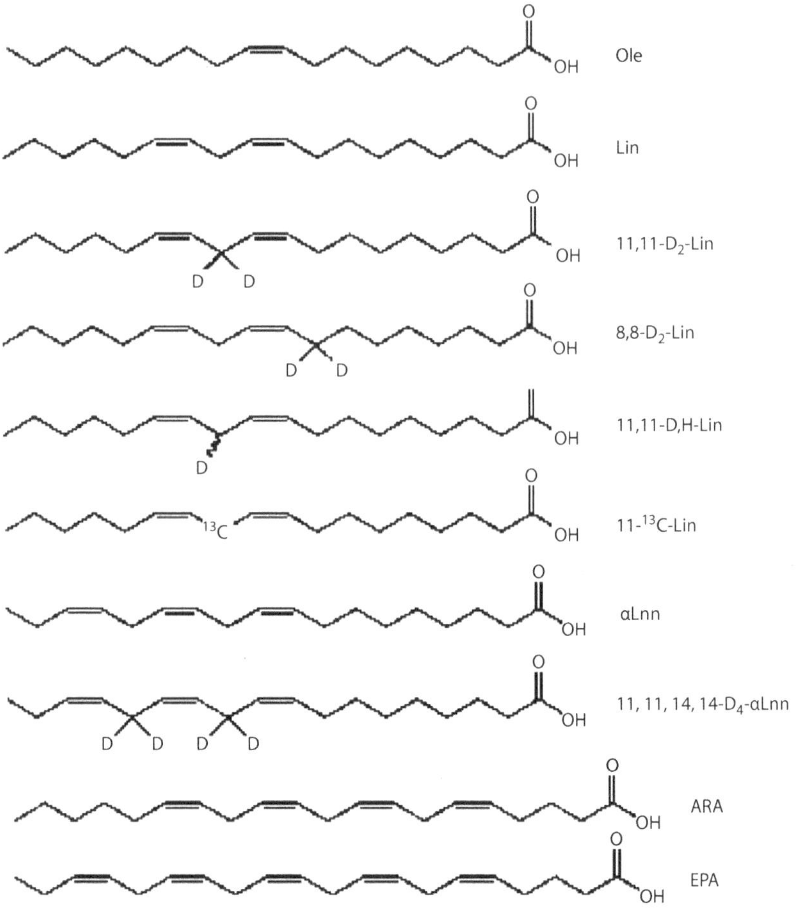

FIGURE 20.1 A battery of deuterium isotopes of fatty acids are being tested for protection of membranes against oxidative damage. (a) Structures of fatty acids used or being considered for deuterium reinforcement studies in yeast, mice, and eventually humans. Ole, oleic acid (18:1, cis-9-octadecenoic acid); Lin, linoleic acid (18:2, cis,cis-9,12-octadecenoic acid); 11,11-D_2-Lin (11,11-D_2-18:2; 11,11-D_2-cis,cis-9,12-octadecenoic acid); 8,8-D_2-Lin (8,8-D_2-18:2; 8,8-D_2-cis,cis-9,12-octadecenoic acid); 11,11-D,H-Lin (11,11-D,H-18:2; 11,11-D,H-cis,cis-9,12-octadecenoic acid); 11-^{13}C-Lin (11-^{13}C-18:2; 11-^{13}C-cis,cis-9,12-octadecenoic acid); αLnn, linolenic acid (18:3, cis,cis,cis-9,12,15-octadecenoic acid); 11,11,14,14-D_4-αLnn (D_4-18:3; 11,11,14,14-D_4-cis,cis,cis-9,12,15-octadecenoic acid); ARA, arachidonic acid; EPA, eicosapentaenoic acid. (From Hill et al., *Free Radic. Biol. Med.* 53:893–906, 2012. Copyright © 2012. Reprinted with permission from Elsevier.) *(continued)*

accelerates death of parkinsonian neurons (see Chapter 3). In the substantia nigra the number of nigral dopaminergic neurons following MPTP exposure in deuterium (D)-PUFA-fed mice is 79.5 percent of the control versus 58.5 percent of the control in (H)-PUFA-fed mice. Biochemical studies of dopamine levels in brain tissue of D-PUFA- versus H-PUFA-fed mice show significant protection by D-PUFA

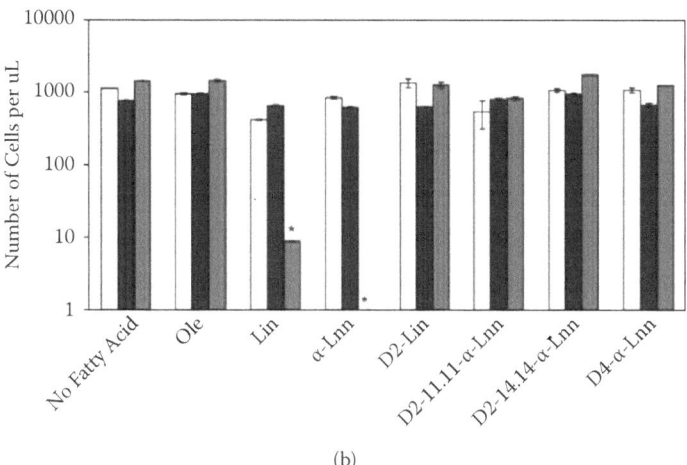

(b)

FIGURE 20.1 *(continued)* A battery of deuterium isotopes of fatty acids are being tested for protection of membranes against oxidative damage. (b) Hill and coworkers (2011) reported a definitive experiment showing that oxidative damage to double bonds of polyunsaturated membranes kills yeast cells. Yeast cells (i.e., coq3 mutants) normally killed by alpha-linolenic acid or linoleic acid become remarkably resistant to toxicity when these membrane building blocks are deuterated. This effect is attributed to the isotope effect of deuterium replacing the facile hydrogen at the double bond, blocking fatty acid peroxidation. The three strains in each set are: wild type (first in panel), atp2 (second), and coq3 (third). (From Hill et al., *Free Radic. Biol. Med.* 50:130–38, 2011. Copyright © 2011. Reprinted with permission from Elsevier.)

of dopamine biosynthesis capacity. Retrotope researchers conclude that dietary D-PUFA partially protects against nigrostriatal damage from oxidative injury elicited by MPTP in mice. Thus, protection of PUFAs against peroxidation by introducing deuterium at the site of double bonds appears to be a universal mechanism explained by the isotope effect. From a fundamental perspective, these data show that oxidative stress can originate at the level of double bonds of unsaturated membrane fatty acid chains. These data have important implications in understanding and treating cardiolipin diseases. Note that the above data do not show directly that cardiolipin is being protected by the deuterium effect. However, in combining data from Retrotope with data discussed next, it seems likely that some of the deuterated fatty acid is incorporated into cardiolipin and protects it against oxidation.

20.4 NEW GENERATION OF ANTIOXIDANTS TARGETING MITOCHONDRIAL MEMBRANES SUPPRESS OXIDATIVE DAMAGE AND IMPROVE MITOCHONDRIAL FUNCTION IN A MOUSE MODEL OF HUNTINGTON'S DISEASE (HD)

XJB-5-131 represents a powerful new generation of mitochondria-targeted antioxidants that suppress oxidative damage to mitochondria (Xun et al., 2012). This bifunctional antioxidant features a mitochondrial membrane-targeting moiety conjugated to a

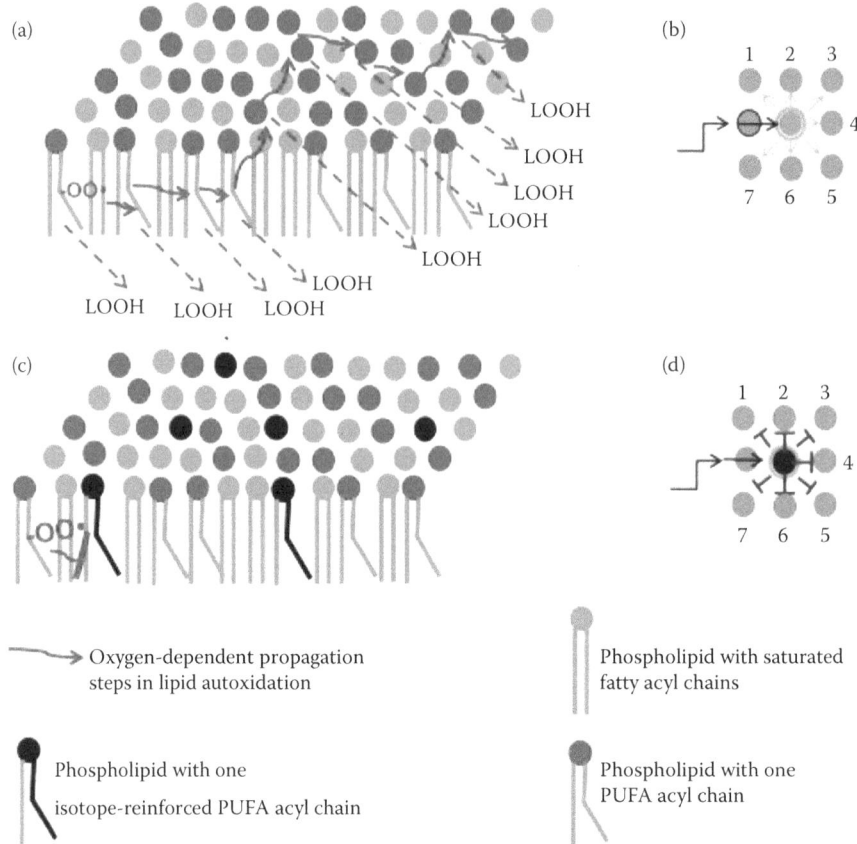

FIGURE 20.2 **See color insert.** Model to explain how deuterium-reinforced PUFAs limit the chain reaction of membrane lipid peroxidation. (a) A theoretical chain reaction is depicted in a membrane where a single initiation event producing a lipid peroxyl radical (denoted by –OO•) starts a chain reaction of lipid autoxidation that in the presence of O_2 may continue indefinitely (red arrows) and produce many molecules of lipid peroxides. Susceptible phospholipid molecules containing a PUFA acyl chain are designated by a kinked blue line and a red dot. (b) Propagation of PUFA autoxidation can progress by interaction with any neighboring PUFAs. (c) The presence of 20 percent isotope-reinforced PUFA (denoted by a black kinked line and a black dot) inhibits (or slows) chain propagation. (d) Propagation is inhibited for PUFAs neighboring the D-PUFA. (From Hill et al., *Free Radic. Biol. Med.* 53:893–906, 2012. Copyright © 2012. Reprinted with permission from Elsevier.)

radical scavenger. The membrane-targeting portion of the molecule is a modification of a peptide segment of the antibiotic gramicidin S that localizes to the mitochondrial membrane. XJB-5-131 might be more efficient than previous mitochondria-targeted antioxidants because its delivery to mitochondria does not depend on the potential gradient across the inner membrane. Instead, localization into the mitochondrial bilayer is directed by the lipophilic peptide segment. As a result, a wide spectrum of mitochondria, including energy-uncoupled mitochondria exhibiting lower membrane potential, is targeted. A second advantage is that the efficacy of a membrane-localized antioxidant may be greater than that of earlier generations of mitochondria-targeted antioxidants that accumulate in the mitochondrial matrix.

Xun and colleagues (2012) show that XJB-5-131 enables suppression of motor decline, inhibits weight loss, reduces mtDNA damage, maintains mtDNA copy number, improves mitochondrial function, and enhances neuronal survival in a mouse model of HD. These data imply that specific targeting of this synthetic antioxidant to mitochondrial membranes has beneficial effects at both the cellular and whole animal level. These findings set the stage for testing of efficacy in a broader set of mitochondrial-mediated diseases and premature aging phenotypes. Xun and coworkers (2012) do not deal with the question of whether a specific molecular species of phospholipid in mitochondrial membranes is being protected by XJB-5-131. However, this topic is covered in a separate paper by Ji and colleagues (2012), discussed next.

20.5 CARDIOLIPIN OXIDATION MEDIATES NEURON DEATH DURING TRAUMATIC BRAIN INJURY IN RATS, AND MITOCHONDRIA-TARGETED ANTIOXIDANT XJB-5-131 PROTECTS CARDIOLIPIN AND PREVENTS APOPTOSIS

A News and Views article by Chan and Di Paolo (2012) in the journal *Nature Neuroscience* describes an important study by Ji and colleagues (2012) concerning the role of cardiolipin as a trigger point for traumatic brain injury (TBI). The general public in the United States has recently been exposed to news reports on TBI as the likely cause of dementia in popular sports figures, especially in professional football and other contact sports, including boxing and soccer. Chan and Di Paolo (2012) point out that in Europe and North America alone, about 3 to 4 million cases of TBI occur, with a fatality rate of 3 to 6 percent. The hidden cost may turn out to surpass that of initial treatment because TBI likely resets and shortens the normal pacemaker of the brain span, resulting in premature onset, by about a decade or two, of dementia. That is, following trauma, TBI is like a ticking time bomb in the brain, giving no visible symptoms or changes that can yet be detected before classic symptoms of neurodegeneration are diagnosed. There is an alarming similarity between TBI-induced events and the effect of aging on brain biochemistry. Following TBI, various forms of neurodegenerative diseases, including Alzheimer's disease, Parkinson's disease, and amyotrophic lateral sclerosis, can emerge early. Even substance abuse and psychiatric disorders can be linked to TBI (Chan and Di Paolo, 2012). Figure 20.3 provides an overview of the molecular pathology associated with TBI that can result from blows or jolts to the head from falls, contact sports, motor vehicle accidents, assaults, and during warfare. Figure 20.3 shows that the primary injury sets in motion a complex cascade(s) of molecular and cellular events that can kill neurons and cause dementia. Figure 20.3 is based on data by Ji and colleagues (2012). The data of these authors support a causal relationship between oxidative damage to cardiolipin and TBI. In essence, CL enriched in the inner mitochondrial membrane acts as a tipping point determining the life or death of neurons after TBI

(a)

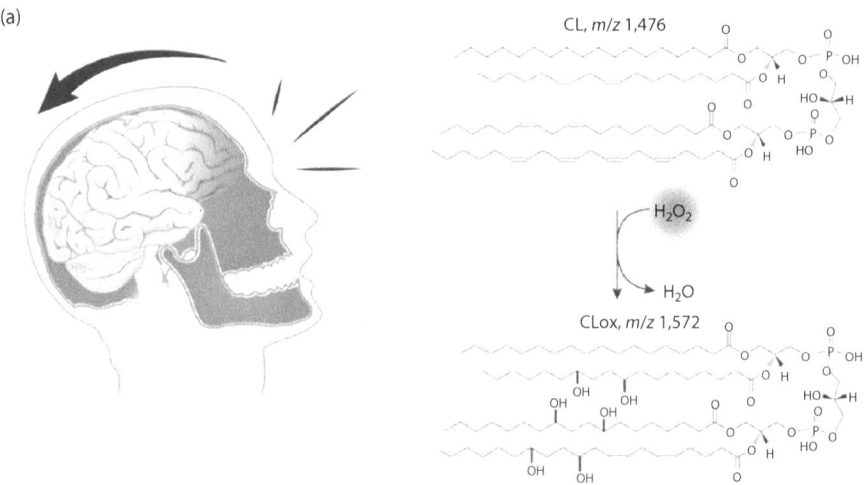

FIGURE 20.3 See color insert. Cardiolipin (CL) as trigger molecule in the molecular pathology of traumatic brain injury (TBI) and discovery of mitochondria-targeted antioxidants protecting mitochondria. (a) Inflammation following TBI causes oxidation of highly unsaturated molecular species of cardiolipin, resulting in dysfunctional derivatives designated CLox. *(continued)*

and suggest a new therapeutic avenue. Once again, the many years that may elapse between the initial TBI and visible pathological events show that the TBI-induced cascade has earmarks of an age-related disease or aging itself.

Ji and coworkers (2012) provide the best evidence yet that cardiolipin, enriched in and being the signature lipid in the inner mitochondrial membrane (Figure 20.3a), is a key mediator of the cascade linking TBI to neuron death. As discussed in previous chapters, the four acyl chains of human CL designated as $(18:2)_4$-CL are characteristic of human mitochondrial membranes, with more highly unsaturated chains being relatively rare. In contrast, CL from many organisms, including rodents, contains polyunsaturated chains ranging from 20:4 (arachidonic acid) to 22:6 (DHA). These data complicate extrapolation from rodents to man. That is, the inner mitochondrial membrane of mice or rats is far more readily oxidized than that of humans, or conversely, human mitochondrial membranes are more saturated and perhaps better shielded against oxidative damage than those of rodents.

The cascade triggering cardiolipin diseases is currently being unraveled at the molecular level. Cytochrome c is part of this cascade and normally behaves as a critical electron carrier shuttling electrons from complex 3 to complex 4 of the mitochondrial electron transport chain. Cardiolipin associates with cytochrome c via negative charges on the CL head group and through hydrophobic forces, effectively keeping cytochrome c anchored to the outer leaflet of the inner membrane. Oxidatively damaged CL has been proposed to trigger the mitochondrial switch from ATP generation to apoptosis initiation. Oxidized CL (CLox) can be generated chemically by peroxidation or by a reaction catalyzed by oxidatively damaged cytochrome c, at which point this electron carrier switches to being a fatty acid peroxidase (Abe

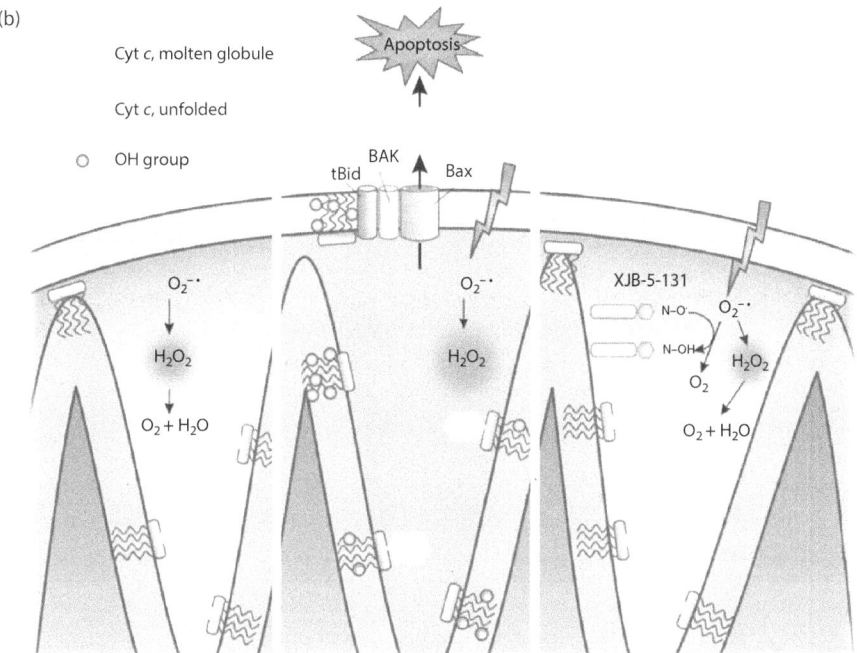

FIGURE 20.3 *(continued)* **See color insert.** Cardiolipin (CL) as trigger molecule in the molecular pathology of traumatic brain injury (TBI) and discovery of mitochondria-targeted antioxidants protecting mitochondria. (b) Left panel shows reactive oxygen species such as H_2O_2 being detoxified and maintained at levels below a critical threshold for initiating a chain reaction. In the middle panel TBI overwhelms defenses against reactive oxygen species, unleashing an oxidative chain reaction propagated by oxidatively damaged CL. Oxidative membrane damage is so severe that membrane-bound cytochrome c is damaged and converted to its peroxidase derivative, further damaging the mitochondrial membrane. Cytochrome c escapes to the cytoplasm, triggering apoptosis. Right panel shows the antioxidant XJB-5-131 being targeted to the inner mitochondrial membrane. This membrane-targeted antioxidant intercepts sufficient numbers of superoxide radical ($O_2^{-\cdot}$), allowing superoxide dismutase to form H_2O_2, which is degraded by catalase yielding O_2 and H_2O. Thus, with the help of the membrane-targeted antioxidant, an oxidative chain reaction leading to cellular death is prevented. Note that the mechanism shown in the right panel can be generalized to cover oxidative stresses generated by aging, neurodegeneration, cancer, chronic inflammation, and mitochondrial diseases as well as TBI. (From Chan and Di Paolo, *Nat. Neurosci.*15:1325–27, 2012. Copyright © 2012. Reprinted by permission from Macmillan Publishers Ltd.)

et al., 2011; Patriarca et al., 2012). Fatty acid peroxidases can be highly active, to the point of producing CLox faster than cellular defensive mechanisms can detoxify this dysfunctional phospholipid. Accumulation of CLox results in translocation of more of this defective lipid from the inner to the outer membrane, where its presence recruits the pro-apoptotic Bcl-2-related proteins tBid, BAK, and Bax, opening mitochondrial transition pores. At the same time, oxidation of cardiolipin weakens the association between cytochrome c and CL, releasing cytochrome c from the inner membrane, permitting escape from the mitochondrion through transition pores of the outer membrane. The presence of cytochrome c in the cytoplasm triggers the apoptosis cascade through sequential interaction with apoptosis protease activator protein-1 (Apaf-1) and caspases 9 and 3. This picture of CL's contribution to apoptosis discussed above served as background for the groundbreaking research by Ji and colleagues (2012), who are focused on curing cardiolipin diseases.

Ji and colleagues (2012) used advanced lipidomics methodology (Samhan-Arias et al., 2012) to show that CL was subject to oxidation, resulting in the generation of 150 different degradation products, including chain-shortened derivatives. In contrast, conventional phospholipids present in the same membrane, including phosphatidylcholine and phosphatidylethanolamine, showed little damage, consistent with the view that CL in mitochondria of rodents has extraordinary sensitivity to oxidation. The amount of CLox increased about twenty-fold, while a similar amount of native CL as substrate for oxidation was consumed. The data confirm that acute brain injury caused by TBI activates CL oxidation. These data open a new window for screening drugs that, after passing across the blood-brain barrier, are able to target brain mitochondria, prevent cardiolipin oxidation, and relieve disease symptoms (Figure 20.3b, right).

Above all, the paper of Ji and colleagues (2012) highlights the role of oxidative damage of cardiolipin in triggering apoptosis and the importance of mitochondria-targeted antioxidants in blocking oxidation of CL. However, this paper does not cover the important linkage between energy stress and aging. Their data do provide clues that energy stress might be involved. TBI-induced peroxidation of CL results in 150 different oxidative products, some of which likely remain in the membrane as dysfunctional lipids. As discussed in Chapters 8 and 13 and in previous books (Valentine and Valentine, 2009, 2013), data generated using chemically defined lipid vesicles and from yeast show clearly that asymmetrical phospholipids such as chain-shortened derivatives of CL oxidation can act directly as potent energy uncouplers of proton gradients. Yeast, for example, naturally produces chain-shortened, saturated phospholipids during anaerobic growth for incorporation and fluidization of its membranes. These phospholipids feature a range of short acyl chains, from C-8 to C-12, paired with longer 18:0 chains in phospholipids, and when incorporated into mitochondrial membranes, they act as uncouplers of proton electrochemical gradients. Also, chemically defined lipid vesicles containing phospholipids formed with acyl chains of C-12 fatty acids are highly permeable to protons, in contrast to lipid vesicles formed from 18:0 and other long-chain fatty acids. Recent data suggest that truncated or chain-shortened membrane phospholipids are major oxidation products formed during TBI-induced peroxidation of CL. We propose that these chain-shortened phospholipids of CL act as energy uncouplers in human mitochondria, supporting the view that oxidative stress and energy stress act in synergy. Both of these powerful stresses might be prevented by membrane-targeted antioxidants.

20.6 SUMMARY

In the 1950s Denham Harman visualized the chemical power of oxidative chain reactions occurring in polyunsaturated membranes as a cause of aging. Later in his career, Harman applied free radical theory in his attempt to understand the role of oxidative chemistry as a possible cause of Alzheimer's disease. Now more than half a century later, two independent research groups report that targeting antioxidants to polyunsaturated mitochondrial membranes of rodents prevents neuron death, with major implications not only for preventing neurodegeneration, but also for treating harmful effects of aging and age-related diseases, including cancer.

Obviously, there are many hurdles to be overcome before discoveries related to protecting cardiolipin in rodent models can be applied in human medicine. Recall that each day billions of human cells, especially epithelial cells such as those lining the colon, naturally senesce to be replaced by a new layer of cells. If senescing cells are not dispatched cleanly, colon polyps can arise and eventually cause colon cancer. An increasing amount of data show that cardiolipin acts as a natural assassin for dispatching lingering colon cells. Thus, the risks of cardiolipin discussed above become a benefit in preventing colon cancer. This benefit might be lost if peroxidation is hindered by drugs such as the antioxidants described here. However, the simple addition of fish oil in the diet might ensure that cardiolipin maintains its potency as a natural oxidative killer of senescing colon cells (see Valentine and Valentine, 2009, for overview and references). The blood-brain barrier represents another major hurdle in developing mitochondria-targeted antioxidants for treating neurodegeneration, by limiting entry of drugs. Interestingly, both deuterated fatty acids and the experimental antioxidant used by Ji and colleagues (2012) exhibit hydrophobic properties facilitating their passage across the blood-brain barrier. Another concern is that the highly unsaturated molecular species of cardiolipin of rodents is atypical when compared to more saturated species of CL found in mitochondrial membranes of humans. A model developed by Hill and coworkers (2012) (Figure 20.2) shows how human $(18:2)_4$-CL might ignite an oxidative chain reaction, spreading damage to the cell. Data by Roginsky (2010) support the model proposed by Hill and colleagues (2012) and show that the rate of peroxidation of $(18:2)_4$-CL is about double that of the methyl ester of 18:2. Also, the rate of $(18:2)_4$-CL oxidation and subsequent pathological effects would be expected to skyrocket after cytochrome c is converted to its peroxidase form in oxidatively damaged mitochondrial membranes.

It is now clear that cardiolipin contributes at least two essential benefits to the cell. The first involves bioenergetic gain, and the second is beneficial pruning of cells during growth and development. In a strange twist of nature, the benefits of cardiolipin are balanced against a potentially lethal risk in which the beneficial role of CL in pruning cells is hijacked, resulting in disease pathologies—cardiolipin diseases. It is safe to say that many secrets of cardiolipin structure and function remain to be discovered.

Ending on a positive note, the availability of several classes of promising mitochondria-targeted antioxidants represents a major milestone toward understanding and treatment of age-dependent diseases and ailments of aging. Ji and colleagues (2012) suggest that windows for drug intervention in the case of TBI might be split, first at the acute phase and later during the long incubation period leading to premature dementia. This opens up the option that one class of targeted antioxidants might be effective when administered immediately after TBI, but only for a short time period. In contrast, a separate membrane antioxidant might be developed when prolonged treatment is required. Ultimately, a cocktail of mitochondria-targeted antioxidants might be developed against a wide range of currently untreatable diseases. We suggest that cardiolipin-targeted antioxidants show enough promise, underpinned by solid science, to be considered a backup to the widely acclaimed human trials involving humanized antibodies that remove beta-amyloid plaque from the brain. Data are expected in the next five years from current human trials testing the amyloid hypothesis. Many neuroscientists are eagerly anticipating the results of these trials, the results of which will determine whether new research and treatment strategies might be needed.

REFERENCES

Abe, M., R. Niibayashi, S. Koubori, et al. 2011. Molecular mechanisms for the induction of peroxidase activity of the cytochrome c-cardiolipin complex. *Biochemistry* 50:8383–91.

Calkins, M. J., M. Manczak, and P. H. Reddy. 2012. Mitochondria-targeted antioxidant SS31 prevents amyloid beta-induced mitochondrial abnormalities and synaptic degeneration in Alzheimer's disease. *Pharmaceuticals (Basel)* 5:1103–19.

Chan, R. B., and G. Di Paolo. 2012. Knockout punch: cardiolipin oxidation in trauma. *Nat. Neurosci.* 15:1325–27.

Do, T. Q., J. R. Schultz, and C. F. Clarke. 1996. Enhanced sensitivity of ubiquinone-deficient mutants of *Saccharomyces cerevisiae* to products of autoxidized polyunsaturated fatty acids. *Proc. Natl. Acad. Sci. USA* 93:7534–39.

Glickman, M. H., and J. P. Klinman. 1995. Nature of rate-limiting steps in the soybean lipoxygenase-1 reaction. *Biochemistry* 34:14077–92.

Hill, S., K. Hirano, V. W. Shmanai, et al. 2011. Isotope-reinforced polyunsaturated fatty acids protect yeast cells from oxidative stress. *Free Radic. Biol. Med.* 50:130–38.

Hill, S., C. R. Lamberson, L. Xu, et al. 2012. Small amounts of isotope-reinforced polyunsaturated fatty acids suppress lipid autoxidation. *Free Radic. Biol. Med.* 53:893–906.

Ji, J., A. E. Kline, A. Amoscato, et al. 2012. Lipidomics identifies cardiolipin oxidation as a mitochondrial target for redox therapy of brain injury. *Nat. Neurosci.* 15:1407–13.

Patriarca, A., F. Polticelli, M. C. Piro, et al. 2012. Conversion of cytochrome c into a peroxidase: inhibitory mechanisms and implication for neurodegenerative diseases. *Arch. Biochem. Biophys.* 522:62–69.

Pryde, K. R., and J. Hirst. 2011. Superoxide is produced by the reduced flavin in mitochondrial complex I: a single, unified mechanism that applies during both forward and reverse electron transfer. *J. Biol. Chem.* 286:18056–65.

Roginsky, V. 2010. Oxidizability of cardiac cardiolipin in Triton X-100 micelles as determined by using a Clark electrode. *Chem. Phys. Lipids* 163:127–30.

Samhan-Arias, A. K., J. Ji, O. M. Demidova, et al. 2012. Oxidized phospholipids as biomarkers of tissue and cell damage with a focus on cardiolipin. *Biochim. Biophys. Acta* 1818:2413–23.

Shchepinov, M. S. 2007. Reactive oxygen species, isotope effect, essential nutrients, and enhanced longevity. *Rejuvenation Res.* 10:47–59.

Shchepinov, M. S., V. P. Chou, E. Pollock, et al. 2011. Isotopic reinforcement of essential polyunsaturated fatty acids diminishes nigrostriatal degeneration in a mouse model of Parkinson's disease. *Toxicol. Lett.* 207:97–103.

Smith, R. A., and M. P. Murphy. 2011. Mitochondria-targeted antioxidants as therapies. *Discov Med.* 11:106–14.

Valentine, R. C., and D. L. Valentine. 2009. *Omega-3 fatty acids and the DHA principle.* Boca Raton, FL: Taylor and Francis Group.

Valentine, R. C., and D. L. Valentine. 2013. *Neurons and the DHA principle.* Boca Raton, FL: Taylor and Francis Group.

Walenga, R. W., and W. E. M. Lands. 1975. Effectiveness of various unsaturated fatty acids in supporting growth and respiration in *Saccharomyces cerevisiae. J. Biol. Chem.* 250:9121–9.

Xun, Z., S. Rivera-Sánchez, S. Ayala-Peña, et al. 2012. Targeting of XJB-5-131 to mitochondria suppresses oxidative DNA damage and motor decline in a mouse model of Huntington's disease. *Cell Rep.* 2:1137–42.

Index

A

Adenine nucleotide translocase (ANT), 167–168
AGE-1, 45
Age-dependent diseases, 1–2, 80–81, 201–202, *See also* Mitochondrial defects or diseases; Neurodegenerative diseases
  cardiolipin hypothesis, 201–202
  Global Burden of Disease Study, 1–2, 201
  mitochondria-targeted antioxidants and treating, 211–212
Akinete, 164
Alpers' syndrome, 90
Altitude adaptations, 65
Alzheimer's disease, 1, 8–9, 80–81, 142–143, 196–197, 207
Ames test for mutagenicity, 20–21
Amyloid-beta plaque, 142–143
Amyotrophic lateral sclerosis, 207
Anaerobic pathway of DHA biosynthesis, 22–23, 27
Animal size and life span, 7
Antioxidants
  anti-aging hypothesis, 4
  mitochondria-targeted, 201–202, 205–207, 210, 211–212
  retina-protective mechanisms, 128
Ants, 57, 75
Apical plasma membrane, 149
Apolipoprotein D (APOD), 194–197
Apoptosis
  cardiolipin oxidation and, 207–210
  hyperactive microglia and, 143
  McIntyre's model, 143–144
  oxidatively damaged cardiolipin and, 208–210
  oxidatively truncated phospholipids and, 141–144
  retinal ganglial cells, 86–87, 133
  truncated phospholipids and mitochondrial mediation, 139, 143–144
Arachidonic acid (ARA), 37, 129, 195–196
Archaea, 29, 112
Ashkenazi centenarians, 189–190
ATP synthase, 96
Autistic tendencies, 79–80
Autophagy, 183

B

Bacillus stearothermophilius, 30
Bacterial models, 19, 27
  anaerobic DHA biosynthesis pathway, 22–23, 27
  applications to mitochondria, 29–31
  DHA detoxification mechanisms, 153
  DHA/EPA excess toxicity, 38
  DHA-/EPA-producing marine bacteria, 21–24
  EPA-minus mutant, 37–38
  EPA recombinants, 24–26
  membrane cation leakiness and, 28
  mutagenicity of reactive oxygen species, 20–21
  oxygen avoidance mechanisms, 163–164
  pathogen membrane-protective mechanisms, 117–118
  sodium ion bioenergetics, 21–22, 28–29
  stabilization of mutants, 31
Bacteroids, 163–164
Baleen whales, 63, 70–72, 79, 154
Barth's syndrome, 88–89, 178
Bee queens, 53–57
Beta-oxidation, 147–148, 151–153, 159
Biosphere distribution of organisms, 26–27
Birds, 63–66
Blood-brain barrier, 148, 152, 153, 210, 211
Bowhead whales, 70–72, 154
Brain
  DHA levels, naked mole rats vs. mice, 70
  DHA turnover, 153
  FOXO factors and longevity, 193
  Harman's hypothesis of aging, 4
  as master regulator of aging, 157–158
    longevity genes, 194–197
  mutator effects and neurodegeneration, 98–100
  oxygen avoidance mechanisms, 161–163
  pigeon-rat life span comparison, 67
  risks of aging, 80–81
Brown adipose tissue, 168, 197

C

Caenorhabditis elegans, 11, 37–51
  dauer stage, 45
  dietary DHA retroconversion to EPA, 37, 38, 41

dietary EPA effects on lifespan, 38
EPA and extreme membrane motion, 40–43
EPA decrease in long-lived mutant, 45–47
highly unsaturated membrane, 37, 39–40
membrane-protective sensory systems, 43–45
mitochondrial cardiolipin levels, 47
nuclear hormone receptors and life span
    regulation, 48–49
peroxidation-resistant methyl-branched fatty
    acids, 48
sensory perception, 42–43
Caffeine, 56
Cancer, 118, 211
Cardiolipin (CL), 59–60, 211
    (18:2)₄-CL, 59, 88, 178, 211
    age-dependent diseases hypothesis, 201–202
    Barth's syndrome and, 88–89
    biosynthesis, 178
    C. elegans membrane protection
        mechanisms, 47
    effect on mitochondrial membrane motion
        and electron transport, 109
    environmental factors and Drosophila
        mitochondria, 59–60
    essential cellular benefits, 211
    mitochondrial membrane composition, 147,
        150–151, 177
    oxidation and neuronal apoptosis, 207–210
    peroxidation sensitivity compared to bacteria,
        19
    traumatic brain injury and, 207–210
Catalase knockout mutant bacteria, 117
Cation leakage, DHA/EPA-enriched membranes
    and, 28
CD36, 141–143
Chain-shortened phospholipids, See Oxidatively
    truncated phospholipids
Charcot-Marie-Tooth disease, 100
Chemosensory neurons, 42–43
"Chicken or egg" question of aging, 3, 8, 92,
    143, 185
Chitin, 72
Chloroplasts, 12, 16–17, 108
Cholesterol, 28, 149, 163–164
CISD2, 98
Colon cancer, 211
Colwellia, 21
Complex 1, See also NADH dehydrogenase
    mitochondrial genes, 96, 101
    nitrogen starvation and oxidative stress, 185
    oxygen concentration and, 160–161
    ROS generation and, 118–122
    ubiquinone deficiency and oxidative stress,
        202–203
Complex 3, 109
Complex 4, 160–161
Convergent evolution of longevity, 79

Cristae membranes, 159–160, See also
    Mitochondrial membranes
Cultural artifacts, 77
Cyanobacteria, 164
Cyclopropane protective groups, 19
Cytochrome c, 208–210, 211
Cytochrome oxidase, 160–161

D

Dauer stage of C. elegans, 45
Deuterated linoleic acid and polyunsaturated
    lipids, 202–205
DFOXO, 60
DHA principle, 8, 26, 150, See also
    Docosahexaenoic acid
    bacteria data and, 22–24, 26
    biosphere distribution of organisms, 26–27
    brain as aging pacemaker, 157
    generalization for polyunsaturated
        membranes, 26
Di-DHA, 40, 126
DNA polymerase, 97, 98–99
Docosahexaenoic acid (DHA), 7
    anaerobic biosynthesis pathway in marine
        bacteria, 22–23
    benefits and risks balance, 8, See also DHA
        principle
    benefits for membrane function, 27, 108–109
    beta-oxidation, 147–148, 151–153, 159
    biosynthesis vs. dietary sources, 153
    comparative biochemistry of detoxification,
        153–154
    deep-sea bacteria, 21–24
    degradation in sperm cells, 4
    di-DHA, 40, 126
    distribution in biosphere, 26
    double bonds and oxidation rate, 43
    Harman's hypothesis of aging, 3–4
    heart levels and life spans, 4, 7, 70
    mammalian membranes and high dietary
        levels, 38
    membrane cation leakiness and, 28
    mitochondrial proton leakage and, 112
    mouse mitochondrial composition, 87
    mouse vs. human membrane content, 150–151
    naked mole rat and mouse life span
        comparison, 69–70
    overexpression and electron transport rates,
        109, 120
    pigeon-rat life span comparison, 66–67
    red blood cell membrane levels, 38
    resistance to cold temperature hardening, 108
    retroconversion to EPA, 37, 38, 41
    rhodopsin membrane disks, 4, 125–127, 159,
        See also Rhodopsin membrane disks
    selective membrane targeting, 148–151

sperm tail enrichment, 129–131, 147, 148
synaptic vesicles and, 149
toxicity in bacteria, 38
tripartite fatty acid blending code, 8–9
truncated oxidation products, 142–143
turnover in brain, 153, 157–159
whale digestive system and, 72, 154
Dominant optic atrophy, 100
Double bonds
C. elegans mitochondrial membranes, 37
membrane peroxidation, 7, 43
terrestrial vs. aquatic insect membrane
composition, 53
Drosophila
apolipoprotein D expression and life span,
194–195
dFOXO and aging, 60
environmental factors and life span, 59–60,
65
NADH dehydrogenase expression and
longevity, 120
POLG gene, 91
royalactin effects, 54
Dual-energy pacemaker model, 8, 193

E

Ecological distribution of organisms, 26–27
Eicosanoids, 196
Eicosapentaenoic acid (EPA), 39
aquatic insect membranes, 53
benefits for membrane function, 27
Caenorhabditis elegans
dietary effects on lifespan and growth, 38
enabling extreme membrane motion,
40–42
membrane content, 37, 39–40, 44
mitochondrial cardiolipin levels, 47
reduction in long-lived mutant membrane,
45–47
sensory perception and extreme
membrane motion, 42–43
distribution in biosphere, 26
double bonds and oxidation rate, 43
mammalian membranes and high dietary
levels, 38
overexpression and electron transport rates,
109, 120
red blood cell membrane levels, 38
retroconversion of dietary DHA, 37, 38, 41
toxicity in bacteria, 38
Electron transport chain
cardiolipin diseases and, 208
chloroplasts and paraquat toxicity, 12
DHA/EPA overexpression in transgenic mice
mitochondrial membranes, 109, 120

mitochondrial membrane mobility and, 107,
108, 109
oxygen and, 160
Endocytosis of synaptic vesicles, 41
Endoplasmic reticulum (ER), 176–177
Energy cost of membrane protection, 2, 83,
172–173
Energy stress
"chicken or egg" question of aging, 3, 8, 92,
143, 185
cost of longevity, 78
dual-energy pacemaker model, 8, 193
dysfunctional mitochondria and, 181, See also
Mitochondrial defects or diseases
hummingbird extreme flight and, 63–65
Leber's hereditary optic neuropathy, 86
McIntyre's model of apoptosis, 143–144
membrane oxidative damage and, 3
mitochondrial hypothesis of aging, 31
mitochondrial proton leakage and, 112
mtDNA mutations and, 101
revised mitochondrial hypothesis of aging,
20, 97–98
Enzymatic detoxification of oxidatively truncated
phospholipids, 144
EPA-minus mutant bacteria, 37–38
EPA recombinant bacteria, 24–26, 27, 31
Epidermal growth factor receptor (EGFr), 54
Epithelial membrane domains, 129, 148–149
Escherichia coli, 19
DHA/EPA excess toxicity, 38
EPA recombinants, 24–26
genetic stabilization, 31
Evolution
convergent evolution of longevity, 79
evidence for human longevity, 76–77
mitochondrial energy efficiency and, 30
oxygen and, 35

F

F-4 neuroprostanes, 158
Fast-firing bending muscles, 40–42
Fatty acid beta-oxidation, 147–148, 151–153, 159
Fenton reagent, 121–122
Ferredoxin, 12
Fish oil components, 41, 195
Flavin mononucleotide (FMN) site, 118–120,
202–203
Food supply and human longevity, 77–78
Fossilized teeth, 76–77
FOXO transcription factors, 45, 78, 190–194
Drosophila dFOXO, 60
Fractalkine, 143
Frankia sp., 163–164
Frataxin, 91–92

Free radicals, *See* Reactive oxygen species
Friedreich's ataxia (FRDA), 91–92
Futile proton cycling, 28–30

G

Genetics and human longevity, 78, 189–198, *See also* Longevity genes
GLaz, 195
Global Burden of Disease Study, 1–2, 201
Glucosamine, 72
Glycosphingolipids, 49
Gramicidin A, 28
Grandparents and human longevity, 75–80
Gray matter, 162–163
GTPases, 99–100

H

Harman, Denham, 3–4
Heart
    DHA levels and life spans, 4, 7, 66–67, 70, 150–151
    Drosophila mitochondrial membranes, 59–60
Heat-producing mitochondria, 168
Helicobacter sp., 117–118
Herbicides, 11–17
High-elevation environments, 65
High-energy electrons, 12–15, 16
Highly unsaturated fatty acid (HUFA)-enriched membranes, *See also* Docosahexaenoic acid; Eicosapentaenoic acid
    aquatic insect membranes, 53
    biosynthesis in naked mole rats, 69
    C. elegans membrane protection mechanisms, 37, 40
    hummingbird extreme flight and, 63–65
    uncoupling protein activation, 171
    whale digestive system and, 72
4-HNE, 131–132
Holy grail of aging, 81, 190
Hummingbirds, 63–65
Huntington's disease, 181, 205–207
Hydrogen peroxide, 117, 118–119, 121–122, 130, 131
Hydroxy fatty acids, 55–56
Hydroxyl radical, 121–122, 131
Hypochlorous acid (HOCl), 118

I

Inflammation, 115, 142–143, 196
Insects, 53, 60–61
    ant diets and longevity, 57
    Drosophila, 54, 59–60
    queen bee longevity, 53–57

Insulin-like growth factor 1 (IGF1), 189–190
Insulin-like regulatory system (ILS), 45, 189
Isotope protection, 16, 202–205

J

Japanese centenarians, 78

L

Lactate homeostasis, 162
Leber's hereditary optic neuropathy (LHON), 85–87, 133
Lens protein, 71
Lethal infantile cardiomyopathy, 88–89
Life span of animals, 70–71
    cardiac DHA levels and, 4, 7, 66–67, 70
    DHA levels and, naked mole rat and mouse comparison, 69–70
    DHA levels and, pigeon-rat comparison, 66–67
    dietary EPA effects in *C. elegans*, 38
    environmental factors and *Drosophila* mitochondrial membranes, 59–60
    hummingbird extreme flight and, 65
    membrane polyunsaturation theory of aging, 7
    membrane unsaturation and queens of social insects, 54
    nuclear hormone receptors and, 48–49
    temperature and, 45, 70, 72
Linoleic acid, 69
    $(18:2)_4$-cardiolipin, 59, 88, 178, 211
    deuterated PUFAs, 202–205
Linolenic acid, 43, 69
Lipid peroxidation, *See* Membrane peroxidation
Lipid peroxyl radical, 130
Lipid raft model, 140
Lipid whiskers, 139–143
Lipocalins, 195
Lipofuscin, 132
LITE-1 photoreceptor, 44
Longevity genes, 78, 81, 189
    apolipoprotein D, 194–197
    Ashkenazi centenarians, 189–190
    FOXO transcription factors, 190–194
    neuronal sources of, 194–197
    screening for, 195
    uncoupling proteins, 197
Longevity in humans
    bioenergetics and, 77–78
    cultural artifacts evidence, 77
    evolution of grandparents, 75–77
    fossilized teeth evidence, 76–77
    genetic mechanisms, 78, 189–198, *See also* Longevity genes
    present-day risks, 80–81

revised holy grail of aging, 81, 190
selective advantage, 75, 150

M

Macrophages (phagocytic cells), 115–118,
  141–143
Marine bacteria, 21–24
McIntyre's model of apoptosis, 143–144
Melanopsin retinal ganglion cells, 133–134
Membrane curvature function, 89, 149
Membrane mobility
  benefits of unsaturated mitochondrial
    membranes, 108–109
  DHA/EPA contribution to, 27, 109
  EPA benefits for *C. elegans*, 40–44
  EPA recombinant bacteria and, 25
  rhodopsin, 107, 108
Membrane pacemaker theory, 7–8, 21, 47, 151
  application to bowhead whales, 72
Membrane peroxidation, *See also* Oxidative
    stress
  bacteria vs. human mitochondria, 19
  C. elegans model, 43–48
  mitochondrial membrane ROS products and,
    121–122
  oxygen concentration and, 160
  protective mechanisms against, *See*
    Membrane-protective mechanisms
  proteins and, 131–133
  rhodopsin membrane disks model, 125–129,
    132–133, *See also* Rhodopsin
    membrane disks
  sperm susceptibility, 130
  theory of aging, 7, 63, 64*f*
    hummingbird extreme flight and, 63–65
    naked mole rat and mouse life span
      comparison, 67–70
    pigeon-rat life span comparison, 66–67
    whale life spans and, 70–72
  truncated phospholipid by-products, 139, 210
  whale digestive system and, 72
  yeast membrane PUFAs and, 202
Membrane polyunsaturation theory of aging,
    7–8, *See also* Membrane peroxidation
Membrane-protective mechanisms, 43, 47, 137
  DHA beta-oxidation, 147–148, 151–153
  DHA selective targeting, 148–151
  energy cost of, 2, 83, 172–173, 191
  evolutionary considerations, 35
  isotope effects, 16, 202–205
  mitochondrial fusion, 101–103
  mitochondrial uncoupling proteins, 78,
    167–173, 194
  mitochondria-targeted antioxidants, 201–202,
    205–207, 210, 211

nematode model (*C. elegans*), 37
  cardiolipin in mitochondrial membranes,
    47
  EPA decrease in long-lived mutant, 45–47
  methyl-branched fatty acids, 48
  sensory systems, 43–45
  oxygen avoidance, 44–45, 131, 157–164
  pathogenic bacteria and, 117–118
  phagocytosis of toxic membranes, 141–142
  queens of social insects, 53–61
  selective fatty acid targeting, 147–155
  whale digestive system and, 72
Membrane-soluble antioxidants, 4
Methyl-branched fatty acids, 48
Mice
  DHA/EPA overexpression and electron
    transport rates, 109
  DHA levels, naked mole rat comparison,
    69–70
  effects of high dietary EPA/DHA levels, 38
  Huntington's disease model, 205–207
  mitochondrial DHA composition, 87
  mitochondrial membrane DHA enrichment,
    150–151
  mitophagy, 185–186
  mutator mouse model, 3, 20, 97–98, 101
  Parkinson's disease model, 203–205
Microglia, 142–143
Mini-mitochondria, 102, 180
Mitochondria-associated membranes (MAMs),
    176–177
Mitochondrial defects or diseases, 85
  Barth's syndrome, 88–89, 178
  excess mitochondrial fission and, 179–181
  Friedreich's ataxia, 91–92
  lactic acid homeostasis and, 162
  late onset, 90–91
  Leber's hereditary optic neuropathy, 85–87,
    133
  miniaturization, 102, 180
  nuclear hormone receptors and, 48–49
  POLG gene mutations, 89–91
Mitochondrial DNA (mtDNA), 20, 83, 85, 95–97,
    125
  excess mitochondrial fission and, 180
  mutations and defects/diseases, 85–93, *See
    also* Mitochondrial defects or diseases
  mutations and errors of replication, 3, 5, 6,
    20, 89, 95, 125
    fusion effects, 99–101
  mutator effects and neurodegeneration,
    98–100
  mutator mouse model, 3, 20, 97–98
  protective effects of fusion, 101–103
  rate of living theory of aging, 63–65
  replication system, 97

Mitochondrial energy output, 95
  aging-associated decline, 2, 3, *See also*
        Mitochondrial hypothesis of aging
  Darwinian selection and, 30
  FOXO factors and longevity, 194
  membrane composition and energy
        conservation balance, 109–113
  uncoupling protein and heat production, 168
Mitochondrial fission, 175
  toxic mitochondria generation, 179–181
Mitochondrial fusion, 99–101
  membrane-protective effects, 101–103
Mitochondrial hypothesis of aging, 2, 3, 4–5, 31
  bacteria data and, 19, 20–21
  historical perspective, 3–4
  hummingbird extreme flight and, 63–65
  membrane polyunsaturation theory of aging,
        7–8
  mutator mouse model, 3
  revised theory, 5–6, 20, 64, 83, 97, 121–122
Mitochondrial infectious damage adaptation
        (MIDA) model, 181
Mitochondrial membranes
  bacterial model applications, 29–31
  benefits of unsaturation, 107–113
  cardiolipin levels, 47, 147, 150–151
  C. elegans model, 37, 47
  DHA benefits/risks balance, 8, 26, 150–151,
        See also DHA principle
  DHA enrichment in mice, 150
  DHA/EPA overexpression and electron
        transport rates, 109, 120
  Drosophila life span and, 59–60
  energy cost of protecting, 2, 83
  fatty acid targeting system, 147
  fusion effects, 99–101
  hummingbird extreme flight and, 63–65
  lactic acid homeostasis, 162
  mitochondria-targeted antioxidants, 201–202,
        205–207, 210, 211–212
  mobility and electron transport, 107, 108, 109
  oxidative damage sensitivity compared to
        bacteria, 19
  oxidatively truncated phospholipids and, 139,
        144, 210
  oxygen avoidance, 159–161
  paraquat toxicity and, 17
  photooxidative damage, 133–134
  proton leakage, 109–113
  reactive oxygen species generation, 121–122
  as targets of oxidation, 125, 131–134
Mitochondrial phospholipids
  molecular species and their biosynthesis,
        177–178
  transport mechanisms, 176–177
Mitochondrial reactive oxygen species (ROS)
        production, 118–122

Mitochondrial uncoupling proteins, *See*
        Uncoupling proteins
Mitochondria origins, 95
Mitochondria-targeted antioxidants, 201–202,
        205–207, 210, 211–212
Mitochondrion functional life span, 175
Mitophagy, 183, 187
  mice and, 185–186
  yeast and, 183–185
Moritella, 21
MPTP, 15–16, 203–205
Mutagenicity, oxygen-damaged membrane fatty
        acids, 20–21
Mutant huntingtin (mHtt), 181
Mutator mice model, 3, 20, 97–98, 101
Myelin and oxygen avoidance, 161–162

N

NADH dehydrogenase, 22, 85, *See also* Complex
        1
  genetic target for mutation, 96
  ROS generation and, 118–120
NADPH, 120
NADPH oxidase, 115–117
Naked mole rats, 63, 67–70, 129–130
Neanderthals, 76, 77
Nematode model, *See Caenorhabditis elegans*
Neurodegenerative diseases, 80–81, 157, *See also*
        specific diseases
  apolipoprotein D and, 196–197
  cardiolipin and traumatic brain injury,
        207–210
  cardiolipin hypothesis of age-dependent
        diseases, 201
  mitochondrial fission and, 181
  mutator gene effects, 98–100
  POLG-mediated mitochondrial diseases,
        89–90
  tripartite membrane fatty acid blending code,
        8–9
Neuronal membrane domains, 149
Niemann-Pick disease, 197
Nitrogen-fixing bacteria, 57, 163–164
Nitrogen starvation and oxidative stress, 185
Nuclear hormone receptors (NHRs), 48–49

O

Oleic acid, 43
Omega-3 fatty acids, 3–4, *See also*
        Docosahexaenoic acid;
        Eicosapentaenoic acid
  fish oil supplement and muscle bending rates,
        41
  naked mole rat dietary requirements, 69
Omega-6 fatty acid, 195

Oxidative bursts, 115–117, 120
Oxidatively truncated phospholipids, 139, 210
    enzymatic detoxification, 144
    lipid whiskers, 139–141
    signaling to phagocytes, 141–142
    sterile inflammation and, 142–143
    triggers of mitochondria-mediated apoptosis,
        143–144
Oxidative stress, 43, *See also* Membrane
        peroxidation; Reactive oxygen species
    bacteria data, 19
    cardiolipin and neuronal apoptosis, 207–210
    C. elegans membrane protection
        mechanisms, 43–47
    "chicken or egg" question of aging, 3, 8, 92,
        143, 185
    dual-energy pacemaker concept, 8
    dysfunctional mitochondria and, 181,
        183–186, *See also* Mitochondrial
        defects or diseases
    environmental factors influencing, 44
    hypothesis of aging, 4–6
        high-energy electrons, 12–15
        membrane polyunsaturation theory of
            aging, 7–8
        paraquat model, 3, 11–17
    Leber's hereditary optic neuropathy and, 87
    McIntyre's model of apoptosis, 143–144
    membrane damage and energy stress, 3
    membrane peroxidation theory of aging, 7, 63,
        64f, *See also* Membrane peroxidation
    mitochondrial contribution, 118–121
    mitochondrial membranes as targets, 125,
        131–134
    nitrogen starvation and, 185
    paraquat and, 3
    retina-protective mechanisms, 128
    sperm cell inactivation, 4
    ubiquinone deficiency and, 202–203
    yeast membrane PUFAs and, 202
Oxygen avoidance, 157
    brain cells and, 161–163
    Caenorhabditis elegans, 44–45
    implications of DHA turnover in brain, 157–159
    mitochondria and, 159–161
    myelin and, 161–162
    root nodule bacteria, 163–164
    sperm cell protective mechanisms, 131, 161
Oxygen deficiency conditions, naked mole rat
        adaptations, 68

P

Palmitic acid, 171
Paraquat, 3, 11–17
    apolipoprotein D expression and resistance to
        toxicity, 195

    C. elegans dauer stage and toxicity resistance,
        45
    high-energy electron transport, 12–15, 17
    human toxicity, 16–17
    neurotoxicity and parkinsonian neurons, 15–16
Parkinsonian neurons, 15–16, 204
Parkinson's disease, 1, 15–16, 197, 203–205, 207
Peroxidation, *See* Membrane peroxidation
Phagocytic cells, 115–118, 141–143
Phagosomes, 129
Phosphatidic acid (PA), 177–178
Phosphatidylcholine (PC), 177–178
Phosphatidylethanolamine (PE), 109, 177–178
Phosphatidylglycerol (PG), 178
Phosphatidylinositol (PI), 42, 177–178
Phosphatidylinositol-3,4,5-triphosphate (PIP$_3$), 45
Phosphatidylserine (PS), 177
Phosphoinositide phosphatase, 41
Phospholipases, 144
Phospholipid transfer in mitochondria, 176–177
Photooxidation, 44, 86–87, 125–127, 133–134
Photosensation in *C. elegans*, 43, 44
Phytoplankton, 26
Pigeons, 63, 66–67
Pigmented epithelial cells, 129
Plastoquinone, 108
POLG gene, 89–91, 97, 98–99
Polyketide pathway of DHA biosynthesis, 22–23
Polyunsaturated fatty acids, *See also* Cardiolipin;
        Docosahexaenoic acid
    balance of benefits and risks for membranes,
        8, *See also* DHA principle
    benefits of unsaturated mitochondrial
        membranes, 107–113
    DHA principle and, 26
    Harman's hypothesis of aging, 4
    membrane oxidative damage and, 3
    membrane polyunsaturation theory of aging,
        7–8, *See also* Membrane peroxidation
    membrane-protective isotope effects,
        202–205
    mitochondrial membrane ROS products and,
        121–122
    oxidatively truncated phospholipids, 144
    photooxidative damage, 133, *See also*
        Photooxidation
    pigeon-rat life span comparison, 66–67
    queen bee longevity and, 56–57
    tripartite fatty acid blending code, 8–9
    yeast membrane susceptibility to oxidative
        stress, 202–203
Protein, dietary modulation and ant longevity, 57
Proteins and membrane peroxidation, 131–133
Proton fidelity, 109–113
Proton leakage, 28–30, 109–113, 149
Proton uncoupling by mitochondria, *See*
        Uncoupling proteins

Q

Queen bees, 53–57
  absent PUFAs and longevity, 56–57
  hydroxy fatty acids and longevity, 55–56
  royal jelly, 54–56

R

Rate of living theory of aging, 63–65
Rat life spans, 66–67
Rattlesnakes, 56
Reactive oxygen species (ROS, or free radicals),
        83, 197
  bacteria data, 19
  bacterial model of mutagenicity, 20–21
  cancer and, 118
  evolutionary considerations, 35
  light damage products, 133
  mitochondrial hypothesis of aging, 2, 3–5, 83
  mitochondrial production, 118–121
    dysfunctional or toxic mitochondria, 181,
        183–186
    membrane products, 121–122
  oxidative bursts, 115–117, 120
  paraquat and generation of, 3, 11–17
  peroxidation, 7, See also Membrane
        peroxidation
  production by phagocytic cells, 115–118
  superoxide, 11, 15, 17, 115, 118–119, 131, 171
  theory of aging, See Oxidative stress
        hypothesis of aging
  uncoupling protein activation, 171, 197
Red blood cells
  DHA/EPA levels, 38
  mitophagy and, 183
Respiration-to-glycolysis ratio, 161–162
Retinal ganglion cells, 86–87, 133–134
Retina-protective mechanisms, 128
Rhodopsin membrane disks, 4, 21, 40, 66–67,
        107, 108, 125–129, 132–133, 159
Rod cells, See Rhodopsin membrane disks
Root nodule bacteria, 163–164
Royalactin, 54
Royal jelly, 54–56

S

Saccharomyces cerevisiae, 202
Salinity and DHA- or EPA-producing bacteria,
        21–22, 25
Salmonella, 19, 20, 21
Saturated fatty acids, uncoupling protein
        activation, 171
Sensory systems, See also Rhodopsin membrane
        disks

C. elegans
  EPA-enabled extreme membrane motion,
        42–43, 44
  membrane protection mechanisms, 43–45
  TRP receptor proteins, 56
Shewanella, 31–32, 37–38, 153
Singlet oxygen, 133
Social insect queens, See Bee queens
Sodium ion transport, 21–22, 28–29, 112
Sperm, 148
  DHA degradation, 4
  DHA-enriched tails, 129–131, 147, 148
  modulation of respiration to glycolysis ratio,
        161
  naked mole rat defects, 69
Sphingolipids, 49
Stearic acid, 49, 142
Sterile inflammation, 115, 142–143
Superoxide dismutase, 117
Superoxide radical, 11, 15, 17, 115, 118–119, 131,
        171, See also Reactive oxygen species
Synaptic vesicles, 41, 149
Synaptojanin, 41

T

Taffazin, 59, 88, 178
Teeth, 76–77
Temperature
  bee longevity and, 57
  Caenorhabditis elegans longevity and, 45
  DHA-/EPA-producing bacteria and, 21, 27
  DHA resistance to hardening, 108
  Drosophila life span and, 59–60, 65
  EPA effects on bending muscle motion in C.
        elegans, 41
  EPA recombinant bacteria and, 25, 31–32
  EPA synthesis in C. elegans and, 39
  high-elevation adaptations, 65
  naked mole rat adaptations, 68, 130
  proton leakage and bacterial adaptations, 30
  rhodopsin disk decay and, 128
  whale life spans and, 70, 72
Termites, 57
Toxic membrane phagocytosis, 141–142
Toxic mitochondria, 185
  excessive mitochondrial fission and, 179–181
  mitophagy and, 183, 184–186
Traumatic brain injury (TBI), 207–210, 211
Tripartite membrane fatty acid blending code, 8
  bacteria data and, 24–29
  likely universality, 27–29
TRPA1, 56
TRP receptor proteins, 56
Truncated phospholipids, See Oxidatively
        truncated phospholipids

U

Ubiquinone, 108, 109
Ubiquinone-deficient mutants, 202–203
UCP1, 168–171
UCP2, 167–171, 197
UCP3, 197
UCP4, 197
Uncoupling proteins (UCPs), 78, 167–173, 194
    activation, 171
    aging and, 172, 197
    longevity genes, 197

V

Vampire bats, 56
Vision-protective mechanisms, 128–129, *See also*
         Rhodopsin membrane disks
Vitamin E, 128

W

Water-wires, 28–29
Whales, 7, 63, 70–72, 79, 154
White matter, 161–163

X

XJB-5-131, 205–207

Y

Yeast, 176, 183–185, 202–203, 210

Z

Zooplankton, 71, 72